T0297041

LONDON MATHEMATICAL SOCIETY STUDENT TEXTS

Managing editor: Professor J. W. Bruce,
Department of Mathematics, University of Hull, UK

3 Local fields, J. W. S. CASSELS
4 An introduction to twistor theory: Second edition, S. A. HUGGETT & K. P. TOD
5 Introduction to general relativity, L. P. HUGHSTON & K. P. TOD
8 Summing and nuclear norms in Banach space theory, G. J. O. JAMESON
9 Automorphisms of surfaces after Nielsen and Thurston, A. CASSON & S. BLEILER
11 Spacetime and singularities, G. NABER
12 Undergraduate algebraic geometry, MILES REID
13 An introduction to Hankel operators, J. R. PARTINGTON
15 Presentations of groups: Second edition, D. L. JOHNSON
17 Aspects of quantum field theory in curved spacetime, S. A. FULLING
18 Braids and coverings: selected topics, VAGN LUNDSGAARD HANSEN
20 Communication theory, C. M. GOLDIE & R. G. E. PINCH
21 Representations of finite groups of Lie type, FRANCOIS DIGNE & JEAN MICHEL
22 Designs, graphs, codes, and their links, P. J. CAMERON & J. H. VAN LINT
23 Complex algebraic curves, FRANCES KIRWAN
24 Lectures on elliptic curves, J. W. S. CASSELS
26 An introduction to the theory of L-functions and Eisenstein series, H. HIDA
27 Hilbert Space: compact operators and the trace theorem, J. R. RETHERFORD
28 Potential theory in the complex plane, T. RANSFORD
29 Undergraduate commutative algebra, M. REID
31 The Laplacian on a Riemannian manifold, S. ROSENBERG
32 Lectures on Lie groups and Lie algebras, R. CARTER, G. SEGAL & I. MACDONALD
33 A primer of algebraic D-modules, S. C. COUNTINHO
34 Complex algebraic surfaces, A. BEAUVILLE
35 Young tableaux, W. FULTON
37 A mathematical introduction to wavelets, P. WOJTASZCZYK
38 Harmonic maps, loop groups, and integrable systems, M. GUEST
39 Set theory for the working mathematician, K. CIESIELSKI
40 Ergodic theory and dynamical systems, M. POLLICOTT & M. YURI
41 The algorithmic resolution of diophantine equations, N. P. SMART
42 Equilibrium states in ergodic theory, G. KELLER
43 Fourier analysis on finite groups and applications, AUDREY TERRAS
44 Classical invariant theory, PETER J. OLVER
45 Permutation groups, P. J. CAMERON
47 Introductory lectures on rings and modules, J. BEACHY
48 Set theory, A HAJNÁL & P. HAMBURGER
49 K-theory for C^*-algebras, M. RORDAM, F. LARSEN & N. LAUSTSEN
50 A brief guide to algebraic number theory, H. P. F. SWINNERTON-DYER
51 Steps in commutative algebra: Second edition, R. Y. SHARP
52 Finite Markov chains and algorithmic applications, O. HAGGSTROM
53 The prime number theorem, G. J. O. JAMESON
54 Topics in graph automorphisms and reconstruction, J. LAURI & R. SCAPELLATO
55 Elementary number theory, group theory, and Ramanujan graphs, G. DAVIDOFF,
    P. SARNAK & A. VALETTE
56 Logic, Induction and Sets, T. FORSTER
57 Introduction to Banach Algebras and Harmonic Analysis, H. G. DALES *et al.*
58 Computational Algebraic Geometry, HAL SCHENCK
59 Frobenius Algebras and 2-D Topological Quantum Field Theories, J. KOCK
60 Linear Operators and Linear Systems, J. R. PARTINGTON
61 An Introduction to Noncommutative Noetherian Rings, K. R. GOODEARL &
    R. B. WARFIELD
62 Topics from One Dimensional Dynamics, K. M. BRUCKS & H. BRUIN
63 Singularities of Plane Curves, C. T. C. WALL
64 A Short Course on Banach Space Theory, N. L. CAROTHERS
65 Elements of the Representation Theory of Associative Algebras Volume 1, I. ASSEM,
    A. SKOWRONSKI & D. SIMSON
66 An Introduction to Sieve Methods and Their Applications, A. C. COJOCARU &
    M. R. MURTY
67 Elliptic Functions, V. ARMITAGE & W. F. EBERLEIN

Hyperbolic Geometry from a Local Viewpoint

LINDA KEEN
Lehman College and Graduate Center
City University of New York

NIKOLA LAKIC
Lehman College and Graduate Center
City University of New York

CAMBRIDGE
UNIVERSITY PRESS

University Printing House, Cambridge CB2 8BS, United Kingdom

One Liberty Plaza, 20th Floor, New York, NY 10006, USA

477 Williamstown Road, Port Melbourne, VIC 3207, Australia

314-321, 3rd Floor, Plot 3, Splendor Forum, Jasola District Centre, New Delhi - 110025, India

103 Penang Road, #05-06/07, Visioncrest Commercial, Singapore 238467

Cambridge University Press is part of the University of Cambridge.

It furthers the University's mission by disseminating knowledge in the pursuit of education, learning and research at the highest international levels of excellence.

www.cambridge.org
Information on this title: www.cambridge.org/9780521863605

© L. Keen and N. Lakic 2007

First published 2007

A catalogue record for this publication is available from the British Library

ISBN 978-0-521-86360-5 Hardback
ISBN 978-0-521-68224-4 Paperback

To Our Families:

Jonathan,
Michael, Erica and Eva,

Branko, Milanka,
Liki and Emily,

for their support and
encouragement.

Contents

Introduction

Geometry is the study of spatial relationships, such as the familiar assertion from elementary plane Euclidean geometry that, if two triangles have sides of the same lengths, then they are "congruent." What does congruent mean here? One possibility, which is rather abstract and very much in the spirit of the axiomatic approach usually attributed to Euclid, is to say:

> Call two straight line segments "congruent" if they have the same length. Call two triangles "congruent" if each side of one can be paired with a side of equal length on the other.

> A more concrete way to say this is that one can take the first line segment and move it "rigidly" from wherever it is in the plane to wherever the second line segment is, in such a way that it overlies the second exactly; similarly, one can take the first triangle and move it rigidly so that it overlies the second exactly.

One of the key insights of modern geometry is that the rigid motions are precisely those maps from the plane onto itself that preserve lengths of line segments. The point is that it is just the notion of "length" that counts: all the angles, the area and other stuff follow once you preserve lengths.

The simplest rigid motion of the plane is reflection in a line: that is, pick a line and, for every point off the line, draw the perpendicular to the line through the point and find the point on the other side that is the same distance from the line; points on the line itself are fixed.

Clearly, following one rigid motion by another results in a rigid motion, and any rigid motion can be reversed to get back where you started, so, in the language of modern algebra, the rigid motions form a group under composition generated by reflections. Historically, the term "symmetry" is used in place of "rigid motion," so you will see a large literature on "groups of symmetries."

1

This analysis raises an obvious question:

Is there some other notion of "length" with a corresponding rigid motion group that is also "natural" and that leads to a geometry that is different from Euclid's? What are the analogues of the notions like "polygon", "area" and "interior angle" that these motions preserve?

The short answer is yes, and the long answer is the heart of this book.

The geometry we will study in this book is called *hyperbolic geometry*. It has the same notion of angle as Euclidean geometry and the rigid motions in this geometry preserve angles.

To motivate our development of this geometry, we show that, by adding a single point to the Euclidean plane, we have another kind of symmetry, "reflection" or "inversion" in a circle: pick any point except the center of the circle and draw the ray from the center through the point; the reflected point is that point on the ray whose distance from the center is the square of the radius of the circle divided by the distance of the given point to the center. This symmetry preserves angles but not Euclidean length. The image of a line under an inversion may be a circle and vice versa so that, in a geometry whose group of motions contains inversions, lines and circles are considered the same kind of object.

In Chapter 2, we begin with the unit disk together with the group of motions that take the disk onto itself and preserve angles. Each such motion is a composition of inversions in circles orthogonal to the boundary of the disk. We develop a notion of length, or distance, for which this group is the group of rigid motions. In this geometry, the shortest path between two points in the disk is along a circle passing through the points and orthogonal to the unit circle. Points on the unit circle are at infinite distance from points inside the disk. The disk with this distance function is called the *hyperbolic plane* and the distance function is called the *hyperbolic metric*.

The most natural way to study this geometry is to use the language of complex numbers and some of the theorems about holomorphic functions – that is, complex valued functions of a complex variable that have a complex derivative. For example, a very important theorem from complex analysis, which we will use over and over again, is called the "generalized Schwarz lemma". It says that any holomorphic map of the disk into itself that is not a rigid motion decreases hyperbolic distance.

We will also be interested in studying geometries for domains that are open subsets of the Euclidean plane. One natural geometry is obtained by restricting the usual Euclidean notion of length. This geometry, however, doesn't really capture some of the intrinsic properties of the domain. For example, rigid motions of the plane do not necessarily map the domain to itself; points on the boundary of the domain will be a finite distance from interior points. These domains do admit other geometries, and in particular a hyperbolic geometry. The idea behind this book is to study a set of related geometries we can put on plane domains that are defined in terms of the derivatives of their distance functions called *densities*. These are more general than the metrics in that they enable us to measure the lengths of paths, the areas of triangles etc. Moreover, the metric may be recovered from the density by integrating over paths. These densities we consider are called "conformal densities" because they have the property that angle measurement is the same for all of the geometries they define.

To begin our discussion on conformal densities for arbitrary plane domains, we need standard tools from complex variable theory and topology. We develop these tools in Chapters 3, 4. We state all of the standard results we need, but we prove only those that are most important or whose proofs involve relevant techniques.

In Chapters 5 and 6 we investigate the symmetry groups of the Euclidean and hyperbolic planes in detail. We show how to use topology to identify plane domains with subgroups of rigid motions. Then in Chapter 7 we define the density that determines the hyperbolic geometry of an arbitrary domain and in Chapters 8, 9 and 10 we define generalizations of this density.

The contents of the first seven chapters of this book are, for the most part, part of the standard mathematical lore. Our approach to defining the hyperbolic metric for an arbitrary plane domain, however, lends itself to generalization and in Chapters 8, 9 and 10 we present this generalization. Some of this material has not appeared before, or has only appeared recently.

In Chapters 11, 12 and 13, we turn to applications. We look at iterated function systems from a given plane domain to a subdomain. The characteristics of the limiting behavior of these systems are controlled by the geometry of the domain and subdomain. This material is the subject of ongoing research and contains both previously unpublished results and open problems.

In general it is a very difficult problem to find an explicit formula for the hyperbolic metric of an arbitrary domain. It is possible, however, to get estimates on the metric by using inclusion mappings. We address this problem in Chapters 14 and 15. In Chapter 14, we get estimates on the hyperbolic metric for various domains by applying the generalized Schwarz lemma to the inclusion map of the domain into the twice punctured plane. We also present an equivalent characterization of the hyperbolic metric which gives another method of finding estimates for the metric. Finally, in Chapter 15 we obtain estimates on the hyperbolic metric for domains called uniformly perfect. In general, one can get estimates using inclusion mappings from the disk into the domain but they only work in one direction. The densities of uniformly perfect domains are comparable to the reciprocal of the distance to the boundary and for these we get estimates in both directions. The last chapter is an appendix in which we present a brief survey of elliptic functions.

The exercises throughout the book should be considered as an integral part of the text material because they contain the statements of many things that are used in the text. In the later chapters they also contain open problems.

We envision this book to be used in several different ways. The first author has used Chapters 1, 4 and 5 as the basis of a junior level undergraduate course. The authors have used the first seven chapters as a one semester second year graduate course. They are currently using the remaining chapters for the second semester of a second year graduate course designed to introduce graduate students to potential research problems.

This book grew out of a seminar for graduate students at the Graduate Center of the City University of New York. While we were writing this book, many of the students read and lectured on the material and gave us invaluable feedback. We would like to thank all the participants in the seminar for all their input. The members of the seminar are Orlando Alonso, Anthony Conte, Ross Flek, Frederick Gardiner, Sandra Hayes, Jun Hu, Yunping Jiang, Greg Markowsky, Bill Quattromani, Kourosh Tavakoli, Donald Taylor, Shenglan Yuan and Zhe Wang. We would also like to thank Ross Flek for making most of the figures in the book and Fred Gardiner for his encouragement. The book would also never have come to being without the strong support of our families, Jonathan Brezin and Ljiljana and Emily Lakic.

1

Elementary transformations of the Euclidean plane and the Riemann sphere

1.1 The Euclidean metric

In most of this book we will be interested in the hyperbolic metric and its generalizations. The hyperbolic metric has its own inner beauty, but it also serves as an important tool in the study of many different areas of mathematics and other sciences.

Before we start exploring hyperbolic geometry in Chapter 2, we get some orientation from the geometry that everyone is very familiar with, Euclidean geometry. We denote the plane by $\mathbb{R}^2$ or $\mathbb{C}$ and a point in the plane either by its Cartesian coordinates, (x, y), by its polar coordinates (r, θ) or by the complex number $z = x + iy = re^{i\theta}$, depending on which is most convenient. We denote the modulus $|z|$ by r and the argument $\arg z$ by θ where $|z| = r = \sqrt{x^2 + y^2}$ and $\arg z = \theta = \arctan(y/x)$. It is also convenient sometimes to think of the point z as the vector from the origin to z.

The argument θ is the angle measured from the positive x-axis to the vector z. The plane has an *orientation*: the argument is positive if the direction from the x-axis to z is counterclockwise and negative otherwise.

The complex conjugate of z is $\bar{z} = x - iy$; it has the same modulus as z and its argument has the opposite sign.

The Euclidean length of the vector z is its modulus and the distance between points $z_1 = x_1 + iy_1$ and $z_2 = x_2 + iy_2$ in the plane is the modulus of their difference,

$$d(z_1, z_2) = |z_2 - z_1| = \sqrt{(x_2 - x_1)^2 + (y_2 - y_1)^2}.$$

This definition of distance satisfies the three requisite conditions for a distance function:

- $d(z_1, z_2) \geq 0$ with equality if and only if $z_1 = z_2$;

5

- $d(z_1, z_2) = d(z_2, z_1)$;
- $d(z_1, z_3) \leq d(z_1, z_2) + d(z_2, z_3)$.

Any space X, together with a distance function d_X satisfying these properties, is called a *metric space*.

Definition 1.1 *A* finite curve *in a metric space is a continuous map of the unit interval* $[0, 1]$ *into a metric space; an* infinite curve *is a continuous map of the real line* $(-\infty, \infty)$ *into a metric space; a* semi-infinite curve *is a continuous map of the half line* $[0, \infty)$ *into a metric space. The word* curve *denotes any one of these.*

We can use the third condition for the distance function to characterize straight lines in Euclidean geometry.

Definition 1.2 *We say a curve is a* straight line *or* geodesic *in the Euclidean plane if for every triple of points* z_1, z_3, z_2 *on the curve with* z_3 *between* z_1 *and* z_2 *we have*

$$d(z_1, z_2) = d(z_1, z_3) + d(z_3, z_2).$$

If the curve satisfying this condition is finite we call it a straight line segment *or* geodesic segment; *if it is a semi-infinite curve satisfying the condition it is called a* ray *or* geodesic ray *and if it is infinite it is called an* infinite geodesic *or an* extended line.

For readability, when it is clear what kind of straight line or geodesic we mean we simply call it a line or geodesic.

Exercise 1.1 Verify that the function $d(z_1, z_2)$ is a distance function on $\mathbb{C}$ and also on any subdomain $\Omega \subset \mathbb{C}$.

Exercise 1.2 Let $z_1 = 1$ and $z_2 = i$. Evaluate the distance from z_1 to z_2 and find the formula for the geodesic segment joining z_1 and z_2.

Exercise 1.3 Show that, for any point z and any extended line l, there is a unique point w on l closest to z. The point w is called the projection from z onto l.

1.2 Rigid motions

The Euclidean geometry of the plane is defined by the following maps of the plane onto itself.

Definition 1.3 A rigid motion *of the plane is a one to one map* f *of the plane onto itself such that, for any two points* z_1, z_2,

$$|f(z_1) - f(z_2)| = |z_1 - z_2|.$$

A *rigid motion is also called an* isometry.

Compositions of rigid motions are again rigid motions and, since they are one to one and onto, they are invertible. We can therefore talk about the *group of rigid motions of the plane.*

The *reflections* in the x- and y-axes respectively are rigid motions given by the maps $R_x : z \mapsto \bar{z}$ and $R_y : z \mapsto -\bar{z}$ respectively. More generally, let l be any line in the plane $\mathbb{C}$. For any point $z \in \mathbb{C}$, there is a point w on l closest to z, the projection from z to l (see Exercise 1.3). Define the *reflection* R_l to be the map that sends a point z to the point $R_l(z)$ on the line through z and w on the opposite side of l such that $d(z, w) = d(R_l(z), w)$. Note that R_l sends a point on l to itself. Clearly a reflection is its own inverse; that is $R_l R_l$ is the identity map Id. Any transformation that is its own inverse is called an *involution.*

Two other types of rigid motions are *translations* given by maps of the form $T_{z_0} : z \mapsto z_0 + z$ and *rotations* given by maps of the form $R_{z_0, \alpha} : z \mapsto z_0 + (z - z_0)e^{i\alpha}$.

We state the properties of rigid motions as Exercises 1.4 to 1.12.

Exercise 1.4 If l_1 and l_2 are lines in $\mathbb{C}$ such that the angle from l_1 to l_2 is α, then the angle from $R_x(l_1)$ to $R_x(l_2)$ is $-\alpha$ and similarly for the images under R_y. In other words, these reflections reverse orientation.

Exercise 1.5 Prove the proposition: A rigid motion is uniquely determined by what it does to three points that do not lie in the same line. Hint: First prove that, if a rigid motion fixes two points, it fixes every point on the line joining them.

Exercise 1.6 Prove the proposition: If l and m are two lines in the plane and R_l and R_m are reflections in these lines then the composition $R_m R_l$ is a rotation or a translation depending on whether the lines intersect or are parallel.

Exercise 1.7 Prove the proposition: If l and m are two lines in the plane and R_l and R_m are reflections in these lines then the composition $R_m R_l R_m$ is a reflection about the line $l' = R_m(l)$.

Exercise 1.8 Prove the proposition: Translations preserve orientation.

Exercise 1.9 Prove the proposition: Rotations preserve orientation.

Exercise 1.10 Prove the proposition: Reflections reverse orientation. Hint: Show that every reflection can be written as a conjugation of the reflection R_x by a translation and rotation.

Exercise 1.11 Prove the proposition: If a rigid motion has a single fixed point it is a rotation; if it fixes two points, but does not fix every point, it is a reflection; if it has no fixed points it is either a translation or a translation followed by a reflection.

Definition 1.4 *A rigid motion that is a translation followed by a reflection is called a* glide reflection. *If the reflection is in a line perpendicular to the vector of translation, the motion reduces to a reflection and the glide reflection is trivial.*

Exercise 1.12 Prove the proposition: Every rigid motion can be written as the composition of at most three reflections. The only rigid motion that cannot be written as a composition of fewer than three reflections is a non-trivial glide reflection. Hint: Write the motion as a composition of reflections and use the fact that reflections about any line are involutions.

Exercise 1.13 Prove the proposition: The group of translations and rotations constitute the full group of orientation preserving rigid motions of the plane.

1.2.1 Scaling maps

In the previous part of this section we looked at the rigid maps of the plane. These maps preserve sizes and shapes. Another map that preserves shapes but changes sizes is the scaling map; that is, contraction or stretching by the same amount in every direction. This map is described by the formula

$$S_c(z) = cz, c(\neq 0) \in \mathbb{C}.$$

The inverse of the scaling map is clearly $S_{1/c}$. Note that the scaling map preserves orientation.

The scaling map can be composed with any rigid motion and the result is also a map that preserves shapes. Such a map is called a *similarity* and the set of such maps forms the group of *similarities* that contains the rigid motions as a subgroup. The subgroup of the similarities that contains all maps of the form $z_0 + cz$, where $z_0, c(\neq 0) \in \mathbb{C}$, is called the *group of complex affine maps* or more simply the group of affine maps.

Exercise 1.14 Show that the similarities take straight lines to straight lines and preserve or reverse the angles between lines.

Exercise 1.15 Show that the affine maps are precisely the subgroup of orientation preserving similarites.

Exercise 1.16 Let $g(z) = 1 + 6z$ be an affine map. What is the image of the unit disk under this map?

1.3 Conformal mappings

Complex affine maps are not the only orientation preserving maps of the plane that preserve angles. Since lines are not necessarily mapped to lines, we need to define what we mean by preservation of angles.

Definition 1.5 *A plane domain Ω is an open connected subset of the complex plane $\mathbb{C}$.*

Definition 1.6 *If two differentiable curves γ_1 and γ_2 in the domain Ω intersect at the point z_0, the angle between them is defined as the angle measured from the tangent to γ_1 to the tangent to γ_2.*

We then define

Definition 1.7 *A differentiable map f from a plane domain Ω in $\mathbb{C}$ to another plane domain X is called conformal at $p \in \Omega$ if, for every pair of differentiable curves γ_1 and γ_2 intersecting at p, the angle between γ_1 and γ_2 is equal to the the angle between the curves $f(\gamma_1)$ and $f(\gamma_2)$ at $f(p)$. If the angle between the curves $f(\gamma_1)$ and $f(\gamma_2)$ at $f(p)$ is the negative of the angle between γ_1 and γ_2 then f is called anti-conformal at p.*

In this chapter we will work with a special class of maps.

Definition 1.8 *An invertible map f from a plane domain Ω in $\mathbb{C}$ onto another plane domain X is called a homeomorphism if both f and its inverse f^{-1} are continuous.*

Definition 1.9 *A homeomorphism f from a plane domain Ω in $\mathbb{C}$ onto another plane domain X is called a diffeomorphism if both f and its inverse f^{-1} are differentiable.*

Definition 1.10 *A diffeomorphism f from a plane domain* Ω *in* $\mathbb{C}$ *to another plane domain X is called* conformal *(or a* conformal homeomorphism) *if it is conformal at every* $p \in \Omega$. *It is called* anti-conformal *if it is anti-conformal at every* $p \in \Omega$.

To find a formula to check whether a map is conformal or anti-conformal we need to use coordinates. We will get formulas in both Cartesian and complex coordinates.

Using Cartesian coordinates we denote the points in Ω by (x, y) and points in X by (u, v) and write $f(x, y) = (u(x, y), v(x, y))$.

It is an exercise from calculus that, if all the directional derivatives at a point have the same magnitude, the map preserves the magnitudes of the angles at that point; that is, the map on tangent vectors is a scaling map. In particular, if we write $f(x, y) = u(x, y) + iv(x, y)$, and use the notation $f_x, f_y,$ u_x, u_y and v_x, v_y for partial derivatives, this means that

$$f_x = u_x + iv_x = -if_y = -iu_y + v_y.$$

Equating real and imaginary parts we get

$$u_x = v_y \text{ and } u_y = -v_x.$$

These equations are called the *Cauchy–Riemann equations* and they are satisfied whenever the map is conformal.

Since we require that f be invertible and orientation preserving, the Jacobian $J(f) = u_x v_y - v_x u_y > 0$. If the orientation is reversed $J(f) < 0$. The condition for anti-conformality is then given by $f_x = if_y$ and we get equations

$$u_x = -v_y \text{ and } u_y = v_x.$$

If we use differentials, the tangents at (x, y) are given by (dx, dy) and the tangents at $(u(x, y), v(x, y))$ are given by (du, dv) and we have

$$du = u_x dx + u_y dy \text{ and } dv = v_x dx + v_y dy.$$

The differential of the map f is

$$df = f_x dx + f_y dy.$$

There is another formal notation that makes these computations neater. Set $z = x + iy, \bar{z} = x - iy, dz = dx + idy, d\bar{z} = dx - idy$ and $f(z) = f(x, y)$. Then write

$$f_z(z) = \frac{1}{2}(f_x(z) - if_y(z)) \text{ and}$$

$$f_{\bar{z}}(z) = \frac{1}{2}(f_x(z) + if_y(z)).$$

The differential of the map becomes

$$df = f_z dz + f_{\bar{z}} d\bar{z} = f_x dx + f_y dy$$

and the Jacobian is $J(f) = |f_z|^2 - |f_{\bar{z}}|^2$. The condition for conformality is $J(f) > 0$, $f_z \neq 0$ and $f_{\bar{z}} = 0$ and the condition for anti-conformality is $J(f) < 0$, $f_{\bar{z}} \neq 0$ and $f_z = 0$. Thus, if f is conformal, $df = f_z dz$ and we can write $f'(z)$ for the derivative of f.

The condition $f_{\bar{z}}(z) = 0$ is the complex *Cauchy–Riemann* equation.

Exercise 1.17 Prove that translations and rotations are conformal maps of $\mathbb{C}$ to itself. Prove that reflections and glide reflections are anti-conformal maps of $\mathbb{C}$ to itself.

Exercise 1.18 Prove that the *real affine* maps defined by $f(x, y) = (x_0 + ax, y_0 + by)$ where a and b are real and neither is zero are conformal if and only if $a = b$.

Exercise 1.19 Prove that the *complex affine maps* defined by $f(z) = z_0 + cz$ are conformal.

In this book we will use the term affine map to mean complex affine map. If the map is a real affine map we will say so.

Exercise 1.20 Set $c = a + ib$, $z_0 = x_0 + iy_0$ and rewrite the map $f(z) = z_0 + cz$ in terms of real coordinates (x, y). Show that there are an angle α and a map $g(x, y) = (x \cos \alpha - y \sin \alpha, x \sin \alpha + y \cos \alpha)$ such that

$$h(x, y) = f \circ g(x, y) = (x_0 + |c|x, y_0 + |c|y).$$

Is the map $g(x, y)$ conformal?

Exercise 1.21 Let

$$\Omega = \left\{ z = x + iy \ \Big| \ -\frac{\pi}{2} < x < \frac{\pi}{2} \right\} \text{ and } X = \mathbb{C} \setminus ((-\infty, -1] \cup [1, \infty)).$$

Show that $f(z) = \sin z$ is a conformal map from Ω onto X.

1.4 The Riemann sphere

The simple map $f(z) = 1/z$ is conformal everywhere it is defined: that is, at every point of $\mathbb{C}$ except the origin. It is very useful to add a single point to the complex plane $\mathbb{C}$ called *the point at infinity* and make it the image of the origin under this map. This can be done so that the resulting map is conformal at the origin and is called a *compactification* of the plane (see Section 4.1 for more about compact spaces). We do it as follows.

Consider the z-plane in Euclidean 3-space with coordinates (x_1, x_2, x_3) where x_1 is identified with the x-axis and x_2 is identified with the y-axis. The standard Euclidean distance between two points $p = (x_1, x_2, x_3)$ and $q = (y_1, y_2, y_3)$ is $d(p, q) = \sqrt{(x_1 - y_1)^2 + (x_2 - y_2)^2 + (x_3 - y_3)^2}$ and the unit sphere is the subset $S = \{p \mid d(p, 0) = x_1^2 + x_2^2 + x_3^2 = 1\}$.

Now draw any ray in $\mathbb{R}^3$ that starts at the north pole, that is, the point $N = (0, 0, 1)$. If this ray goes through a point $z \in \mathbb{C}$, it will also intersect the unit sphere at the point

$$x_1 = \frac{z + \bar{z}}{1 + |z|^2}, \quad x_2 = \frac{z - \bar{z}}{i(1 + |z|^2)}, \quad x_3 = \frac{|z|^2 - 1}{|z|^2 + 1}.$$

This is a bijective map between the plane and all points except $(0, 0, 1)$ on the sphere. It is called *stereographic projection*. The inverse map is given by

$$z = \frac{x_1 + i x_2}{1 - x_3}.$$

It is easy to compute that points in the interior of the disk, $\{z \mid |z| < 1\}$, are identified with points on the southern hemisphere $\{x_3 < 0\}$; the south pole $(0, 0, -1)$ corresponds to the origin $z = 0$. Points in the exterior of the disk, $\{z \mid |z| > 1\}$, are identified with points on the northern hemisphere $\{x_3 > 0\} \setminus \{(0, 0, 1)\}$.

It is therefore natural to add a point, "the point at infinity", to the plane as the point identified with $(0, 0, 1)$. We call this extended plane the *Riemann sphere* and denote it by $\overline{\mathbb{C}}$.

The involution $z \mapsto 1/z$ of $\overline{\mathbb{C}}$ corresponds to the involution $(x_1, x_2, x_3) \mapsto (x_1, -x_2, -x_3)$ of the sphere and the involution $z \mapsto 1/\bar{z}$ corresponds to the involution $(x_1, x_2, x_3) \mapsto (x_1, x_2, -x_3)$.

It is obvious that every straight line on the plane transforms into a circle on the sphere passing through the north pole; the circle is the intersection of the plane through the given line passing through the north pole. It is less obvious that every circle on the sphere corresponds to a circle or straight line on the plane. To see this, note that every circle on the sphere is the intersection of a plane $ax_1 + bx_2 + cx_3 = d$ with the sphere. Substituting we get

$$a(z + \bar{z}) - ib(z - \bar{z}) + c(|z|^2 - 1) = d(|z|^2 + 1).$$

In terms of x and y we have

$$(d - c)(x^2 + y^2) - 2ax - 2by + (c + d) = 0.$$

Therefore, if $c \neq d$ the circle on the sphere transforms to a circle on the plane and if $c = d$ it transforms to a straight line.

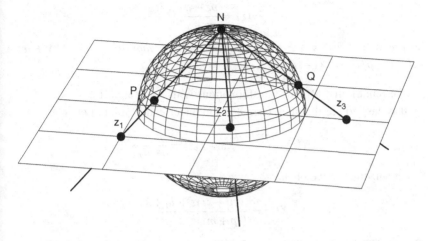

Figure 1.1   Stereographic projection. The points P and Q are on the sphere

Exercise 1.22 There is a natural reflection in the sphere given by the map $(x_1, x_2, x_3) \mapsto (x_1, x_2, -x_3)$ that interchanges the two hemispheres. Show that the map on the corresponding points in the plane is given by the *inversion in the unit circle* defined by the map $z \mapsto 1/\bar{z}$.

Exercise 1.23 Let a be any point in the complex plane and let $R > 0$. The map $z \mapsto a + R^2/(\overline{z - a})$ is called the *inversion in the circle* $|z - a| = R$. Show that it is an anti-conformal involution.

Exercise 1.24 Find the formula for the circle on the unit sphere that corresponds to the circle $\{z \,|\, |z - 1| = 1\}$ under stereographic projection.

1.5 Möbius transformations and the cross ratio

Note that every affine map g is also a conformal map of the Riemann sphere with $g(\infty) = \infty$. On the other hand, if f is any conformal map of the Riemann sphere with $f(\infty) = c \neq \infty$, then the map $g(z) = 1/(f(z) - c)$ is a conformal map of the plane. Letting $g(z)$ be an affine map leads to the following definition.

Definition 1.11 *A* Möbius transformation *of the Riemann sphere* $\overline{\mathbb{C}}$ *is a map of the form*

$$X(z) = \frac{pz+q}{rz+s},$$

where p, q, r *and* s *are complex numbers such that* $ps - qr \neq 0$. *We let* $X(\infty) = p/r$ *and* $X(-s/r) = \infty$.

We remark that Möbius transformations with $r = 0$ are affine. It is easy to see that they form a group under composition. Let

$$X(z) = \frac{pz+q}{rz+s} \text{ and } Y(z) = \frac{jz+k}{lz+m}.$$

Then substituting we obtain

$$Y \circ X(z) = \frac{(jp+kr)z + jq + ks}{(lp+mr)z + lq + ms}.$$

Since $(jp + kr)(lq + ms) - (jq + ks)(lp + mr) = (ps - qr)(jm - lk) \neq 0$, this is again a Möbius transformation. For simplicity we write XY for $X \circ Y$.

Proposition 1.5.1 *The Möbius transformations form a group of conformal maps of the Riemann sphere* $\overline{\mathbb{C}}$.

Proof. We have seen that Möbius transformations are closed under composition.

To find the inverse of Y explicitly, set

$$Z(z) = \frac{mz - k}{-lz + j}.$$

Since $mj - (-k)(-l) = jm - kl \neq 0$, Z is a Möbius transformation. Substituting and canceling we have $YZ(z) = ZY(z) \equiv z$ so that Y is the inverse of Z.

Clearly $X_{\bar{z}} = 0$. Note that, if $r = 0$, $X'(z) = p/s \neq 0$ and $X(z)$ is conformal. If $r \neq 0$, $X'(z) = (ps - qr)/(rz + s)^2$ so that $X'(z) \neq 0$ for any $z \in \mathbb{C}$. For $z = \infty$, we set $w = 1/z$ and evaluate the derivative of $X(1/w)$ at $w = 0$. We get

$$X'(z) \frac{dz}{dw} = \frac{-(ps - qr)}{r^2} \neq 0$$

and conclude that Möbius transformations are conformal homeomorphisms of the Riemann sphere.

The other group axioms follow directly. $\square$

Definition 1.12 *Suppose* $\lambda(\Omega)$ *is a complex number defined for plane domains* Ω *(or* $\overline{\mathbb{C}}$*). If, for every conformal f defined on* Ω, $\lambda(\Omega)$ *satisfies*

$$\lambda(f(\Omega)) = \lambda(\Omega)$$

then λ *is called a* conformal invariant *for* Ω; *that is,* λ *is invariant under conformal maps defined on* Ω.

Note that, since the composition of a conformal map with a Möbius transformation is conformal, any conformal invariant of a subdomain of the Riemann sphere is invariant under Möbius transformations. It is easy to compute that any Möbius transformation is completely determined by its values at three distinct points. Any non-trivial conformal invariant of the sphere therefore must therefore be defined on a subdomain of the Riemann sphere that has at least four points in its complement. A conformal invariant that involves exactly four points is the *cross ratio*.

To define the cross ratio take a quadruple of distinct points a, b, c and d in $\overline{\mathbb{C}}$. Denote the vectors joining a to b, b to c, c to d, and d to a by L, M, R and T respectively; if the points are colinear L, M, R and T stand for left, middle, right and total respectively.

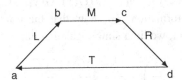

Figure 1.2   The intervals of the cross ratio

Definition 1.13 *The* cross ratio *is defined as*

$$cr(a, b, c, d) = \frac{LR}{MT} = \frac{(b-a)(d-c)}{(c-b)(a-d)}. \tag{1.1}$$

Clearly, the cross ratio is a nonzero complex number. An easy calculation also shows that the cross ratio cannot be equal to 1. Observe that each of the four points a, b, c, d appears once in the denominator and once in the numerator in the formula for the cross ratio but with opposite sign. As an example, let us suppose that all four points a, b, c and d lie on the same line l. Then all the vectors L, R, M and T have the same direction, up to sign, and thus the cross ratio is a real number.

To see the invariance of the cross ratio under Möbius transformations, suppose that a, b, c and d are any four distinct points and let $X(z) = (pz + q)/$

$(rz + s)$ be a Möbius transformation. If $r = 0$, then $X(z) = \frac{p}{s}z + \frac{q}{s}$ is affine and $X(z)$ keeps the cross ratio intact; that is,

$$cr(x, y, z, w) = cr(X(x), X(y), X(z), X(w)).$$

If $r \neq 0$, then $X(z) = BCDEF(z)$, where

$$B(z) = \frac{p}{r} + z,$$

$$C(z) = \frac{qr - ps}{r}z,$$

$$D(z) = \frac{1}{z},$$

$$E(z) = z + s$$

and

$$F(z) = rz.$$

Clearly, the affine transformations B, C, E and F each leave the cross ratio invariant because $cr(x, y, z, w) = cr(G(x), G(y), G(z), G(w))$, whenever G is any affine transformation. To show that the transformation D leaves the cross ratio invariant, we do a simple calculation:

$$cr\left(\frac{1}{x}, \frac{1}{y}, \frac{1}{z}, \frac{1}{w}\right) = \frac{(\frac{1}{y} - \frac{1}{x})(\frac{1}{w} - \frac{1}{z})}{(\frac{1}{z} - \frac{1}{y})(\frac{1}{x} - \frac{1}{w})}$$

$$= \frac{(x - y)(z - w)}{(y - z)(w - x)} = cr(x, y, z, w).$$

Therefore, the cross ratio is an invariant under all Möbius transformations $X(z)$.

We have seen that any four colinear points have real cross ratio. Conversely, suppose that three of the four distinct points a, b, c and d are colinear. Then, obviously, in order for the cross ratio of this quadruple to be real, the fourth point must belong to the same line.

Consider Möbius transformations of the form $A(z) = e^{i\theta}(z + c)/(1 + \bar{c}z)$ where θ is any real number, and c is any point in the unit disk Δ. It is easy to check that if $|z| = 1$ then $|A(z)| = 1$ and $A(0) = ce^{i\theta} \in \Delta$ so that A is a Möbius map of Δ onto itself.

Also consider the Möbius transformation $B(z) = i(1 + z)/(1 - z)$. By calculating that the expression $i(1 + e^{i\theta})/(1 - e^{i\theta})$ is real we verify that $B(z)$

maps the unit circle with the exception of the point 1 onto the real axis. This combined with the invariance of the cross ratio under all Möbius transformations says the following.

Theorem 1.5.1 *Four distinct point a, b, c, d belong to the same line or circle if and only if their cross ratio is real.*

To simplify notation we will slightly change the definition of the word colinear, and say that four distinct points a, b, c, d are *colinear* if they belong to the same line or to the same circle. Since the cross ratio is invariant under Möbius transformations, Theorem 1.5.1 implies that Möbius transformations map colinear points to colinear points.

Exercise 1.25 The Möbius transformation $X(z) = 1/z$ is the composition of the inversion in the unit circle $z \mapsto 1/\bar{z}$ and the reflection in the real axis $z \mapsto \bar{z}$. Show that if I_1 and I_2 are inversions in any pair of circles their composition is always a Möbius transformation. Show also that the inversion I_1 in some circle composed with the reflection in any line is always a Möbius transformation.

Exercise 1.26 Show that any Möbius transformation can be written as the composition of an even number of inversions in circles and reflections in lines. Hint: Use what you know about rigid motions and the above exercise.

Exercise 1.27 Complete the proof of Theorem 1.5.1.

Exercise 1.28 The definition of the cross ratio $X = cr(x, y, z, w)$ depends on the order of the four points. There are 24 permutations of this order. Show that there are only six different values for cross ratio for any permutation:

$$X, \frac{1}{X}, 1 - X, \frac{1}{1-X}, \frac{X-1}{X}, \frac{X}{X-1}.$$

These transformations of X form the dihedral group of order 6.

Exercise 1.29 If a, b, c and x, y, z are any two triples of distinct points in the Riemann sphere, show there exists exactly one Möbius transformation X such that $X(a) = x$, $X(b) = y$ and $X(c) = z$.

Exercise 1.30 Find the Möbius transformation $X(z)$ that sends points $0, 1$ and 2 to points $1, 2i$ and $3i$ respectively.

Exercise 1.31 Use the cross ratio to find the equation of the circle that contains points $1 - i, 1 + i$ and 16.

18		*Elementary transformations*

Exercise 1.32 Let $p = 0$, $q = 3$, $r = 1 + i$ and $s = 2 + i$. Evaluate the cross ratio $cr(p, q, r, s)$ and the cross ratio $cr(f(p), f(q), f(r), f(s))$ where f is the real affine map $f(x, y) = (3x, 1 + 2y)$.

Exercise 1.33 Show that the cross ratio of four distinct points is never equal to 1.

Exercise 1.34 Let $X(z) = z/(z - 3)$. Evaluate the Euclidean center of the disk $X(\Delta)$.

Exercise 1.35 Prove that every Möbius transformation of the unit disk onto itself can be written in the form

$$A(z) = e^{i\theta} \frac{z + a}{1 + \bar{a}z},$$

where θ is a real number, and a is a point in the unit disk.

1.5.1 Classification of Möbius transformations

Let $X(z) = (pz + q)/(rz + s)$ be a Möbius transformation. Since we may divide the numerator and denominator by the same complex number without changing the action of $X(z)$ on $\overline{\mathbb{C}}$, we may assume without loss of generality that $ps - qr = 1$.

We can solve for the fixed points of $X(z)$:

$$z_1, z_2 = \frac{p - s \pm \sqrt{(p + s)^2 - 4}}{2r}.$$

We call $p + s$ the *trace* of $X(z)$ and denote it by $\operatorname{tr} X$. Generically $(p + s)^2 - 4 \neq 0$ and there are two fixed points. We assume this is true for now and return to the case when it is not later.

Given any two points z_1, z_2 in the plane, there is a one parameter family of circles that pass through them. As usual, we consider the line through the points as a "circle through infinity". The centers of the circles in the family vary along the perpendicular bisector of the line joining them; the "center" of the circle through infinity is the point at infinity on the perpendicular bisector. Denote this family by H_{z_1, z_2} and call it the *hyperbolic pencil* of circles determined by z_1, z_2. There is a dual one parameter family of circles orthogonal to the hyperbolic family denoted by E_{z_1, z_2} and called the *elliptic pencil*. It contains a circle through infinity, the perpendicular bisector. The centers of the circles in E_{z_1, z_2} lie on the line through z_1, z_2 outside the segment joining them; the point at infinity on this line is the "center" of the perpendicular bisector.

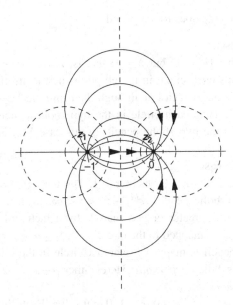

Figure 1.3   The family H_{z_1,z_2} are the solid circles and the family E_{z_1,z_2} are the dotted circles

The Möbius transformation X preserves these pencils. To see this, recall that Theorem 1.5.1 implies that Möbius transformations map colinear points to colinear points. Thus, since X fixes z_1 and z_2 it sends a circle in H_{z_1,z_2} to another circle in H_{z_1,z_2}. Since X preserves orthogonality it sends an orthogonal circle in E_{z_1,z_2} to another orthogonal circle in E_{z_1,z_2}. See Figure 1.3.

It is instructive, however, to look more closely at what happens. Note that, if $A(z) = (z - z_1)/(z - z_2)$ is the Möbius transformation that sends z_1 to 0 and z_2 to ∞, then $A(z)$ maps H_{z_1,z_2} to the pencil $H_{0,\infty}$ and E_{z_1,z_2} to the pencil $E_{0,\infty}$. Whatever X does to one of the circles in the pencil H_{z_1,z_2}, $AXA^{-1}(z)$ does to its image in the pencil $H_{0,\infty}$, and similarly for the elliptic pencils.

From Exercise 1.36 below we see that

$$Y(z) = AXA^{-1}(z) = \lambda z = \frac{\sqrt{\lambda}z + 0}{0z + 1/\sqrt{\lambda}}$$

where $\sqrt{\lambda} + \frac{1}{\sqrt{\lambda}} = p + s$ and that $H_{0,\infty}$ consists of all lines through the origin and $E_{0,\infty}$ consists of all circles whose center is the origin. Note that we may replace the coefficients p, q, r, s by their negatives without changing the map X so that it doesn't matter which square root we choose.

Let us see what $Y(z)$ does to the pencils:

1. **The elliptic case**:
   Assume first that $|\lambda| = 1$, but $\lambda \neq 1$; that is, $\lambda = e^{i\theta}$, $0 < \theta < 2\pi$. Then $Y(z) = e^{i\theta}z$ maps each circle in the elliptic pencil to itself, rotating it by an angle of θ. It maps each line through the origin, $te^{i\alpha}$, $-\infty < t < \infty$, to the line $te^{i(\alpha+\theta)}$. Thus each circle in the hyperbolic pencil is mapped to another circle in the hyperbolic pencil. In this case $Y(z)$ is called *elliptic* and $-2 < p+s < 2$.

2. **The hyperbolic case**:
   Assume next that $\lambda = r > 0$ where $r \neq 1$. The map $Y(z) = rz$ maps each line in the hyperbolic pencil, $te^{i\alpha}$, $-\infty < t < \infty$, to itself, expanding or contracting as r is greater or less than 1, by a factor of r. Each circle $te^{i\alpha}$, $0 \leq \alpha < 2\pi$, is mapped to the circle $tre^{i\alpha}$, $0 \leq \alpha < 2\pi$, so each circle in the elliptic pencil is mapped to another circle in the elliptic pencil. In this case $Y(z)$ is called *hyperbolic*. Here, either $p+s < -2$ or $p+s > 2$.

3. **The loxodromic case**:
   Finally assume $\lambda = re^{i\theta}, r > 0, r \neq 1, 0 < \theta < 2\pi$. Each line in the hyperbolic pencil is expanded (or contracted) and then rotated into another line. Similarly each circle in the elliptic pencil is expanded (or contracted) into another circle in the pencil and it, in turn, is rotated. In this case $Y(z)$ is called *loxodromic*. Note that in this case $p+s$ is not real valued.

To see how $X(z)$ acts on the circles in its hyperbolic and elliptic pencils, we map a circle to its image under A, see what Y does to it, and then map back by A^{-1}. If Y is elliptic, we also call X elliptic and we see that X acts invariantly on each circle of the elliptic pencil E_{z_1,z_2} and invariantly on the whole hyperbolic pencil H_{z_1,z_2}, sending one circle in the pencil to another. Similarly, if Y is hyperbolic, we call X hyperbolic and see that X acts invariantly on each circle of the hyperbolic pencil H_{z_1,z_2} and invariantly on the whole elliptic pencil E_{z_1,z_2}, sending one circle in the pencil to another. Finally, if Y is loxodromic, we also call X loxodromic. In this case, no circle in either pencil is left invariant; rather, any such circle is mapped to another in the same pencil.

To summarize: the two fixed points of $X(z)$ determine two pencils of circles invariant under the action of $X(z)$. The action on the pencils is determined by the *multiplier* λ *of* X defined in terms of the *trace* $p+s$ *of* X.

4. **The parabolic case**:
   Now let us consider the case where the trace $p+s = \pm 2$, the multiplier $\lambda = 1$ and $X(z)$ has only one fixed point z_1. Such an X is called *parabolic*.

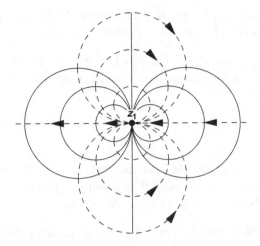

Figure 1.4   The invariant circles of a parabolic transformation

As in Exercise 1.37 below, there is a Möbius transformation $A(z)$ such that $Y(z) = AXA^{-1}(z) = z+1$ and the single fixed point of Y is at infinity. There are again two orthogonal pencils of circles through infinity (straight lines), the horizontal and vertical lines. The pair of pencils is clearly invariant under Y. The horizontal lines are mapped to themselves and each vertical line is mapped to its translate by 1.

We can map the horizontal and vertical lines by A^{-1} to a pair of orthogonal pencils of circles all of which pass through z_1. This pair of pencils is invariant under X. See Figure 1.4.

Remark. Suppose $X^n = Id$ for some n. Then, choosing A as above, depending on whether X is parabolic or not, we also have $Y^n = AX^nA^{-1} = Id$. If X is parabolic, however, $Y^n = z + n$ which is never the identity. Otherwise $Y^n = r^n e^{in\theta} z$. This can only be the identity if $r = 1$ and $\theta = 2m\pi/n$ for some integer m. Thus the only finite order Möbius transformations are elliptic transformations whose multiplier argument is a rational multiple of 2π.

Exercise 1.36 Suppose $X(z) = (pz + q)/(rz + s)$ with $ps - qr = 1$ and $(p + s)^2 \neq 4$. Let z_1, z_2 be its fixed points. Show that, if $A(z) = (z - z_1)/(z - z_2)$, then

$$Y(z) = AXA^{-1}(z) = \lambda z = \frac{\sqrt{\lambda}z + 0}{0z + 1/\sqrt{\lambda}}$$

where $\sqrt{\lambda} + 1/\sqrt{\lambda} = p + s$ and $H_{0,\infty}$ consists of all lines through the origin and $E_{0,\infty}$ consists of all circles whose center is the origin.

Exercise 1.37 Suppose $X(z) = (pz+q)/(rz+s)$, $ps - qr = 1$ and $p+s = 2$. Show that there is a Möbius transformation $A(z)$ such that $Y(z) = AXA^{-1} = z+1$.

Exercise 1.38 Find the orthogonal pencils of circles for the transformation $X(z) = 1/(z+1)$.

Exercise 1.39 Find all Möbius transformations that map the imaginary axis onto the unit circle, send the point 0 to 1, and the point i to -1. Which of those are hyperbolic? parabolic? elliptic? loxodromic?

Exercise 1.40 Show that any circle from the hyperbolic pencil H_{z_1, z_2} intersects any circle from the elliptic pencil E_{z_1, z_2} at exactly two points x and y. Show that $cr(x, z_1, y, z_2) = -1$.

Exercise 1.41 Show that x and y belong to a circle from the hyperbolic pencil H_{z_1, z_2} if and only if $\Im cr(x, z_1, y, z_2) = 0$.

Exercise 1.42 Show that x and y belong to a circle from the elliptic pencil E_{z_1, z_2} if and only if $|cr(x, z_1, y, z_2)| = 1$.

Exercise 1.43 Show that a Möbius transformation $w = X(z)$ that sends z_0 to w_0 and has fixed points z_1, z_2 can be written as

$$\frac{w - z_1}{w - z_2} = \frac{z - z_1}{z - z_2} \, cr(z_1, w_0, z_2, z_0).$$

Also show that, if, for some Möbius transformation B, $BXB^{-1} = \lambda z$, then $\lambda = cr(z_1, w_0, z_2, z_0)$.

1.6 Möbius groups

Let us call the full group of Möbius transformations $\mathcal{M}$ and the subgroup preserving the unit disk Γ. Note that the identity for these groups is the transformation $Id(z) = z$.

There is a natural identification of $\mathcal{M}$ with the group of complex 2×2 matrices with determinant one, modulo sign, $PSL(2, \mathbb{C}) = SL(2, \mathbb{C})/\pm Id$, where

$$\pm \begin{pmatrix} a & b \\ c & d \end{pmatrix} \Longleftrightarrow \frac{az+b}{cz+d}, \quad a, b, c, d \in \mathbb{C}, \ ad - bc = 1.$$

By Exercise 1.35, elements of the subgroup Γ correspond to matrices as follows:

$$\pm \begin{pmatrix} \dfrac{e^{i\theta/2}}{\sqrt{1-|a|^2}} & \dfrac{ae^{i\theta/2}}{\sqrt{1-|a|^2}} \\ \dfrac{e^{-i\theta/2}\bar{a}}{\sqrt{1-|a|^2}} & \dfrac{e^{-i\theta/2}}{\sqrt{1-|a|^2}} \end{pmatrix} \iff e^{i\theta}\,\frac{z+a}{1+\bar{a}z}.$$

The trace of an element in $SL(2,\mathbb{C})$ is defined as $a+d$. It is easily seen to be invariant under conjugation; that is, for any pair of elements $A, B \in SL(2,\mathbb{C})$, the elements A and BAB^{-1} have the same trace. For an element of $\mathcal{M}$ the trace is defined only up to sign.

We will often find it convenient to identify a Möbius transformation with a matrix in $SL(2,\mathbb{C})$ and will tacitly do so. For readability, we will often abuse notation and refer to the trace of an element of $PSL(2,\mathbb{C})$ or $\mathcal{M}$. By this we mean the trace of either lift to $SL(2,\mathbb{C})$. In all but a few exceptional cases, the lift to $SL(2,\mathbb{C})$ from $PSL(2,\mathbb{C})$ is well defined. We will discuss these cases when they arise.

The upper half plane, $\mathbb{H}$, is conformally equivalent to Δ; for example, the Möbius transformation $W = (iz+i)/(-z+1)$ maps Δ onto $\mathbb{H}$. Since the Möbius transformations leaving $\mathbb{H}$ invariant also leave the real axis invariant, it is easy to check that these are of the form

$$A(z) = \frac{az+b}{cz+d}, \quad a, b, c, d \in \mathbb{R}, \ ad - bc = 1.$$

As above, these maps are identified with the group $PSL(2,\mathbb{R})$ consisting of the real 2×2 matrices of determinant one, modulo sign.

The groups Γ and $PSL(2,\mathbb{R})$ are conjugate subgroups in $PSL(2,\mathbb{C})$; that is, every element $B \in PSL(2,\mathbb{R})$ can be written as $B = WAW^{-1}$ for some $A \in \Gamma$ and vice versa. In our geometric investigations in later chapters, we will be interested in the properties of subgroups of Γ and $PSL(2,\mathbb{R})$ that are invariant under conjugation such as the trace. We will therefore often go back and forth between a subgroup of Γ and its conjugate in $PSL(2,\mathbb{R})$, using whichever is more convenient.

Exercise 1.44 Show that if A and B are matrices representing a pair of Möbius transformations then the matrix representing the composed Möbius transformations is the matrix product AB.

Exercise 1.45 Verify that every Möbius transformation of the upper half plane onto itself corresponds to a matrix in $PSL(2,\mathbb{R})$.

Exercise 1.46 Show that if A and B are elements of $PSL(2,\mathbb{C})$ then $\operatorname{tr} A = \operatorname{tr} BAB^{-1}$.

1.7 Discreteness of Möbius groups

We can identify the complex plane with the set of 1×1 complex matrices $Z = (z)$ which is called $GL(1, \mathbb{C})$. This space has a natural norm $||Z|| = |z|$ that makes it into a metric space, and the metric is the usual Euclidean metric (see Section 1.1).

Analogously, the space of all 2×2 complex matrices is denoted by $GL(2, \mathbb{C})$ and it also has a norm: if $A = \begin{pmatrix} a & b \\ c & d \end{pmatrix}$ then

$$||A|| = (|a|^2 + |b|^2 + |c|^2 + |d|^2)^{\frac{1}{2}}.$$

This norm induces a metric $d(A, B) = ||A - B||$ on this space. The subset $SL(2, \mathbb{C}) \subset GL(2, \mathbb{C})$ forms a subspace of this metric space.

The trace is a well defined function from $SL(2, \mathbb{C})$ to $GL(1, \mathbb{C})$ and it is continuous with respect to the metrics induced by the norms. Similarly, the square of the trace is a well defined function from $PSL(2, \mathbb{C})$ to $GL(1, \mathbb{C})$. We leave the proofs of these statements to Exercises 1.47–1.50.

It is obvious that traces of all the elements of $SL(2, \mathbb{R})$ are real. Since trace is invariant under conjugation, the traces of the elements in Γ, which are defined up to sign, are also real.

Definition 1.14 *A subgroup G of $\mathcal{M}$ or $SL(2, \mathbb{C})$ is called* discrete *if there is a neighborhood N of the identity Id in $SL(2, \mathbb{C})$ such that $G \cap N = \{Id\}$.*

Clearly many subgroups of $\mathcal{M}$ are not discrete. For example, let $a_n \to 0$ be a sequence of points in Δ and let

$$A_n = \frac{z + a_n}{1 + \bar{a}_n z}.$$

Note that as $n \to \infty$ the transformations A_n tend to the identity transformation $Id(z) = z$. Suppose G is the smallest subgroup of $\mathcal{M}$ containing all A_n; then clearly G is not discrete. This leads us to give an equivalent statement of the definition of discreteness (see Exercise 1.52):

Proposition 1.7.1 *A subgroup G of $\mathcal{M}$ is discrete if and only if there is no sequence of distinct elements $A_n \in G$ with $A_n \to Id$.*

Discreteness is invariant under conjugation. Precisely,

Proposition 1.7.2 *Suppose $G \subset SL(2, \mathbb{C})$ and let B be any element of $SL(2, \mathbb{C})$. Then G and the group G' consisting of all elements of the form $C = BAB^{-1}$ for all $A \in G$ are either both discrete or both non-discrete.*

Proof. By Proposition 1.7.1, if G is not discrete there is a sequence $A_n \in G$, $A_n \to Id$. By Exercise 1.50 below, the sequence $BA_nB^{-1} \to Id$ also. If G' is not discrete, the same argument applies to $C_n \in G'$ and $B^{-1}C_nB$. $\square$

A corollary of Proposition 1.7.1 and Exercise 1.50 is

Corollary 1.7.1 *If G contains a sequence of distinct elements A_n tending to some element $A_\infty \in SL(2, \mathbb{C})$ it cannot be discrete.*

We can apply this corollary to show

Proposition 1.7.3 *Suppose G contains an elliptic element with multiplier $\lambda = e^{2\pi i\theta}$ for θ irrational. Then G is not discrete.*

Proof. Since θ is irrational, the sequence $a_n = n\theta$ mod 1 has a convergent subsequence $a_{n_j} = n_j\theta$ mod 1 with limit $\phi \in [0, 1]$. It follows that the sequence A^{n_j} is conjugate to $e^{2\pi i n_j\theta}z \to e^{2\pi i\phi}z$ so that A^{n_j} also converges to a limit. Hence G is not discrete. $\square$

Following Poincaré we define

Definition 1.15 *A discrete subgroup of $\mathcal{M}$ or $PSL(2, \mathbb{C})$ is called a* Kleinian group.

and

Definition 1.16 *A discrete subgroup of Γ or $PSL(2, \mathbb{R})$ is called a* Fuchsian group.[1]

In this book, we will be most interested in the geometric properties of Fuchsian groups. The geometry of Kleinian groups is a rich and interesting theory and would need a book of its own. There are several; for example, see [27], [39] and [46].

Since all the traces in a Fuchsian group are real, the elements are classified as elliptic, parabolic or hyperbolic; there are no loxodromic elements in a Fuchsian group.

Exercise 1.47 Show that $||AB|| \leq ||A|| \, ||B||$. Verify that $d(A, B) = ||A - B||$ defines a metric on the space $GL(2, \mathbb{C})$.

Exercise 1.48 Show that

$$\begin{pmatrix} a_n & b_n \\ c_n & d_n \end{pmatrix} \to \begin{pmatrix} a & b \\ c & d \end{pmatrix}$$

---

[1] See Section 6.1 for the history of this nomenclature.

where the convergence is in norm if and only if $a_n \to a$, $b_n \to b$, $c_n \to c$ and $d_n \to d$. It follows that as a metric space $GL(2, \mathbb{C})$ is equivalent to $\mathbb{C}^4$ with the usual Euclidean metric. It also follows that $SL(2, \mathbb{C})$ is equivalent as a metric space to the set $\{(a, b, c, d) \in \mathbb{C}^4 \mid ad - bc = 1\}$.

Exercise 1.49 Prove that the trace function $\operatorname{tr} A = a + d$ is a well defined function continuous with respect to the metric on $SL(2, \mathbb{C})$. Deduce from this that the square of the trace $(\operatorname{tr} A)^2 = (a + d)^2$ is well defined on $PSL(2, \mathbb{C})$ and is also continuous with respect to the induced metric on the quotient.

Exercise 1.50 If $X \in SL(2, \mathbb{C})$ show that the maps $f(A) = AX$ and $g(A) = XA$ are homeomorphisms of $SL(2, \mathbb{C})$ to itself with respect to the metric $d(A, B)$.

Exercise 1.51 If $X \in \mathcal{M}$ show that the maps $f(A) = AX$ and $g(A) = XA$, where, as usual, multiplication here is composition of Möbius transformations, are homeomorphisms of $\mathcal{M}$ to itself with respect to the metric $d(A, B)$.

Exercise 1.52 Prove Proposition 1.7.1.

Exercise 1.53 Show that the fixed points of hyperbolic and parabolic elements of a Fuchsian group lie on $\mathbb{R}$ (or $\partial\Delta$). In addition show that the fixed points of elliptic elements are complex conjugate (inverse with respect to $\partial\Delta$). Thus, R or $\partial\Delta$ belongs to the elliptic pencil of the fixed points of any elliptic element. It belongs to the hyperbolic pencil of the fixed points of any hyperbolic element, and to the invariant pencil of any parabolic element of a Fuchsian group.

Exercise 1.54 Prove that if an element A of a Kleinian or Fuchsian group is elliptic it must have finite order, that is $A^n = Id$ for some integer $n > 0$.

1.8 The Euclidean density

In this section we introduce a new approach to measuring distances: the "local" or "infinitesimal" approach.

Suppose that you are on an island in the shape of a plane domain Ω. You are standing at the spot p and you want to get to some other spot q on the same island by traveling the shortest possible distance. That is, you try to find the shortest path connecting the two points p and q and calculate the Euclidean length of this path. If the straight line segment joining p to q lies inside Ω, then we can use it. This situation is shown in Figure 1.5, where the shortest path from p to q is the line segment and the Euclidean distance is equal to the Euclidean length of that segment; its value is $|p - q|$.

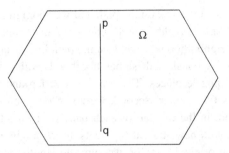

Figure 1.5   The path stays inside the island

There is a name for domains with this property.

Definition 1.17 *A plane domain is called* convex *if any pair of points in the domain can be joined by a line segment completely contained in the domain.*

If, however, the straight segment connecting p to q goes through the ocean, as in Figure 1.6, then we need to look at other paths joining p to q. That is, paths that stay on the island.

In other words, for non-convex domains like the one in Figure 1.6, we need a tool to measure the Euclidean length of any path joining p to q and try to find the shortest one. This tool comes from the "local" or "infinitesimal" picture. If we zoom in far enough on the picture of the whole island Ω to a tiny neighborhood of any point a in Ω, the resulting picture will be a round disk entirely contained in Ω, and any two points in that disk will have a straight line segment connecting them. That straight line segment is the shortest path connecting those points. We think of this neighborhood and its line segment as a piece of a puzzle that we use to find the length of the path from p to q.

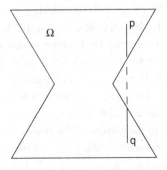

Figure 1.6   The straight path does not stay on the island

First, we pick a set of points on the path and we zoom in on these points to get a collection of small neighborhoods whose union contains the whole path. In each of these neighborhoods we replace the path by a straight line and find its length. That is, we consider these neighborhoods with their line segments as a collection of puzzle pieces. These pieces overlap and the overlaps tell us how the pieces fit together. Second, we zoom out a bit on all these tiny neighborhoods and, in the overlap of each pair, we connect the ends of the straight segments with new straight segments, to get a new path from p to q that is close to the original. Finally, we sum the lengths of the paths in the neighborhoods to find the length of the whole path. The smaller the pieces, the better the approximation is to the length of the path.

More precisely, the standard Euclidean distance $d(t, s) = |t - s|$ that we defined for the whole plane restricts to the infinitesimal neighborhoods and, in the limit, it induces a density $d(t)|dt| = |dt|$ defined at each point t. To find the length of any path from p to q in Ω, we integrate the density $d(t)|dt|$ along the path. This integral measures the Euclidean distance along the path. To find $d_\Omega(p, q)$, the distance between the points, we take the infimum over all paths that join p to q and remain inside the island Ω.

Definition 1.18 *Set*

$$d(\gamma) = \int_\gamma d(t)|dt| \text{ and } d_\Omega(p, q) = \inf_\gamma d(\gamma)$$

where the infimum is over all paths γ in Ω joining p to q. If there is no path for which the integral is finite, we take $d_\Omega(p, q) = \infty$.

Notice that we are using the same notation for both the density and the distance between points. This should not cause any confusion because the number of arguments in the domain for the two notions is different. It should, in fact, enhance readability in the chapters to come, because we will introduce several more densities. Since we will be working with several densities and using the same notation for both the density and the distance, it will be easier to recognize which type of density or distance we are talking about.

Inside any puzzle piece the shortest Euclidean paths are always straight line segments. Since Euclidean distance is translation invariant, the Euclidean density $d(t)|dt|$ satisfies $d(t) = d(t + s)$ and $d(t)$ is a constant function. This constant is of course positive, and it represents the zooming factor that takes us from the picture of Ω to the picture of the puzzle piece. Zooming back out to the picture of Ω, and connecting the pieces of the puzzle, we see that $d(t)$ is a constant function throughout the island Ω. It follows that this density function is determined by its value at a single point. This all becomes very

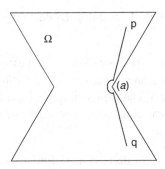

Figure 1.7 $d(p, q) = |p - a| + |a - q|$

important if, for example, we want to make a road atlas of the island Ω, and the island is too big to fit on one page. To make it possible for the reader of the road atlas to clearly see all the details of the island, we use several pages. To make every point and its immediate neighborhood visible on at least one page of the map, we include ample overlaps between pages. The overlaps can be tiled together into one puzzle by using the translation mappings. If we use enough pages we can use the same zoom factor on each puzzle piece; this factor is usually called the scale of the atlas. Therefore, to find the shortest path between any two points on the island, we can use the road atlas, and find all possible paths including those that may go through many different pages of the atlas. Because we use translation mappings to move from one page of the atlas to another, the lengths of these paths are all well defined.

From a global point of view, the scale is the same anywhere on Ω. Thus, to find the length of a path we may take $d(t)$ to be constantly equal to 1 on Ω, and use Definition 1.18. In the case where there is no shortest path connecting p and q, for example in Figure 1.7, the infimum of the definition is not a minimum.

This method also works for connected subsets of higher dimensional Euclidean spaces. For example, let S be the unit sphere in $\mathbb{R}^3$. Suppose we have two points p and q on S and we want to find the shortest path on S that connects them. We apply Definition 1.18 with $d(t) = 1$ and paths γ that lie on the sphere S to obtain the distance

$$d_S(p, q) = \inf_{\gamma} \int_{\gamma} d(t)|dt|.$$

Exercise 1.55 Let p and q be any two points on the unit sphere S. Show that $d_S(p, q)$ is the length of a circular arc on S that goes from p to q and has a center at 0. That is the minimizing arc γ is in the intersection of S with the plane that goes through 0, p and q. Evaluate $d_S((1/\sqrt{3}, 1/\sqrt{3}, 1/\sqrt{3}), (-1, 0, 0))$.

Exercise 1.56 For any two points z and w in the extended plane, let $s(z, w)$ be the distance between z and w defined by projecting z and w to points p and q on the unit sphere S, and then setting $s(z, w) = d_S(p, q)$. The distance $s(z, w)$ is called the *spherical distance between z and w*. Evaluate $s(.85i, 3)$. Is there a spherical density $s(t)$?

Exercise 1.57 Show that $d_{\mathbb{C}}$ defined in Definition 1.18 coincides with the Euclidean density d defined in Section 1.1.

Exercise 1.58 Find a plane domain Ω that is non-convex where d_Ω is the restriction of d to Ω.

Exercise 1.59 Show that a plane domain Ω is convex if and only if the infimum in Definition 1.18 can be replaced by the minimum; that is there is a shortest curve joining any two points in Ω. Furthermore, that shortest curve must be the line segment joining the two points.

Exercise 1.60 Show that a plane domain Ω is convex if and only if, for every pair of points p and q, the set of points r in $\mathbb{C}$ such that $d(p, q) = d(p, r) + d(r, q)$ is a subset of Ω.

This exercise gives a way to extend the notion of convex to subsets of the sphere.

Definition 1.19 *A subset X of the sphere S is called* convex in S *if for every pair of points p, q on S the set of points r on S such that $d_S(p, q) = d_S(p, r) + d_S(r, q)$ belong to X.*

Exercise 1.61 Is the set $X = \{(x_1, x_2, x_3) \mid x_3 > -.2\}$ convex in S?

Definition 1.20 *A plane domain Ω is* star-like *if there is a point $p \in \Omega$ so that for any other point q in Ω the straight line segment joining p to q lies inside Ω.*

Exercise 1.62 Find an example of a star-like domain Ω that is not convex.

Exercise 1.63 Generalize the definition of star-like to subsets of any metric space. Find one subset of S that is star-like and one that is not.

Exercise 1.64 Let Ω be the exterior of the unit disk. Evaluate the Euclidean distance $d_\Omega(3i, -2i)$.

Definition 1.21 *A sequence of points a_n in a space X with a metric $d(x, y)$ is called a Cauchy sequence if, for every $\epsilon > 0$, there is some integer $N = N(\epsilon)$ such that $d(a_m, a_n) < \epsilon$ for all $m, n > N$. X is a complete metric space if every Cauchy sequence converges to a point in X.*

Exercise 1.65 Which plane domains Ω with the metric d_Ω defined by the Euclidean density are complete metric spaces?

Exercise 1.66 Show that $d(t) = 1$ is the infinitesimal form of d_Ω for any domain Ω. That is, show that, for any $x \in \Omega$, we have $d_\Omega(x, x+t) = t d(x) + o(t)$ as $t \to 0$.

Exercise 1.67 Consider the space $GL(2, \mathbb{C})$ with the metric induced by the norm $\|A\|$. Show that the density this metric defines is the same as the Euclidean density coming from the representation of the space as a subset of the Euclidean space $\mathbb{C}^4$. Is the subset $SL(2, \mathbb{C})$ convex? star-like?

Exercise 1.68 Suppose Ω is a convex plane domain. Remove a point from Ω, and call the result Ω^*. Is Ω^* still convex? If p and q are any two points in Ω^* compare $d_\Omega(p, q)$ and $d_{\Omega^*}(p, q)$. Now remove a disk D from Ω and call the result Ω^{**}. Is Ω^{**} still convex? If p and q are any two points in Ω^{**} compare $d_\Omega(p, q)$ and $d_{\Omega^{**}}(p, q)$.

1.8.1 Other Euclidean type densities

In the examples above the density was the same at every point. In some applications, however, the density function is not the same at every point. For example, if we want to know how long it will take to get from one place in our island Ω to another, there may be snow or bad roads that we have to cross so that the time is not directly proportional to the distance. We can reflect this by making the density function different where the snow is or where the bad roads are. Some specific examples of this are in the following exercises.

Exercise 1.69 Let SQ be the square with vertices $-3 - 3i, 3 - 3i, 3 + 3i$ and $-3 + 3i$ and let L be the closed square with vertices $-1 - i, 1 - i, 1 + i$ and $-1 + i$. Define the density $c(z)$ on SQ as follows. Let $c(z) = 16$ for all $z \in L$ and let $c(z) = 1$ for all $z \in SQ \setminus L$. Evaluate the distance $c(-1.5, 0.75i)$.

Is there a shortest path that joins -1.5 to $0.75i$ with respect to the density $c(z)$?

Exercise 1.70 Let Δ be the unit disk in $\mathbb{C}$. Define the density $r(z) = 1/(1 - |z|)$ on Δ. Use the density to find $r(1/3, i/3)$. Is Δ with this metric a complete metric space?

2

Hyperbolic metric in the unit disk

In this chapter we define the hyperbolic plane in terms of the hyperbolic metric for the unit disk. As we will see in Chapter 7, we will use the hyperbolic plane to define a hyperbolic metric on an arbitrary plane domain.

2.1 Definition of the hyperbolic metric in the unit disk

In Chapter 1 we defined Euclidean density. We now use the same approach to find the formula for a hyperbolic density $\rho(t)|dt|$ in the unit disk. Once we have the density, we define the hyperbolic distance as we did in the Euclidean case as the infimum of lengths of paths. In this context a path is a continuous map of the unit interval $[0, 1]$ into the unit disk. As in the Euclidean context, a curve may be either a path or a continuous map of the semi-infinite interval $[0, \infty)$ or the infinite interval $(-\infty, \infty)$ into the disk.

Definition 2.1 *Let γ be any path in the unit disk $\Delta = \{z \mid |z| < 1\}$ joining the points p and q. The ρ-length of γ is defined as*

$$\rho(\gamma) = \int_\gamma \rho(t)|dt|$$

and the ρ-distance from p to q is defined as

$$\rho(p, q) = \inf \rho(\gamma)$$

where the infimum is over all possible paths γ.

To get an explicit formula for the Euclidean density $d(t)$, we used the fact that Euclidean distance is invariant under translations. Analogously, to derive an explicit formula for the hyperbolic density $\rho(t)$, we will require that it be invariant under conformal homeomorphisms of the unit disk. That means

32

that, if we decide to make a road atlas of Δ with several pages (charts), we must use Möbius maps instead of translation maps as overlap maps between pages.

A precise statement that the hyperbolic density $\rho(t)|dt|$ be invariant under all Möbius transformations $A : \Delta \to \Delta$ is

$$\rho(A(t))|A'(t)| = \rho(t). \tag{2.1}$$

We saw in Exercise 1.35 that all such homeomorphisms A are of the form

$$A(z) = e^{i\theta} \frac{z+a}{1+\bar{a}z},$$

where θ is a real number, and a is a point in the unit disk.

Observe that the mapping A moves the point zero to a point a, which may be any point in the unit disk. Therefore, applying formula (2.1) to the transformation $A : \Delta \to \Delta$ above, we see that the hyperbolic density on the unit disk is, like the Euclidean density, determined by its value at a single point. To simplify future notation, we let $\rho(0) = 1$.[1] Then from formula (2.1) we obtain

$$\rho(a) = \frac{1}{|A'(0)|} = \frac{1}{1 - |a|^2}. \tag{2.2}$$

Note that the density at a depends only on its Euclidean distance from the origin.

Exercise 2.1 Show that, if A is any Möbius homeomorphism of Δ and γ is any path in Δ, then $\rho(\gamma) = \rho(A(\gamma))$. Moreover, show that $\rho(p, q) = \rho(A(p), A(q))$ for any two points p and q in Δ.

2.1.1 Hyperbolic geodesics

From the formula for the hyperbolic density and Definition 2.1, we can find the formula for the hyperbolic distance between any two points p and q in the unit disk. Suppose first that one of the two points, say q, is the origin 0. If the point p is also the origin, then, obviously, the distance between p and 0 is zero. Suppose that p is a nonzero point in the unit disk and let γ_0 be the straight segment that joins 0 to p, $\gamma_0(t) = tp, 0 \leq t \leq 1$. By formula (2.2), the hyperbolic length of γ_0 is

$$\rho(\gamma_0) = \int_{\gamma_0} \rho(t)|dt| = \int_0^1 \frac{1}{1 - t^2|p|^2}|p|dt,$$

---

[1] The alternative normalization $\rho(0) = 2$ is also often found in the literature.

and an easy calculation gives us

$$\rho(\gamma_0) = \frac{1}{2} \log \frac{1 + |p|}{1 - |p|}.$$

Now suppose γ is any path joining 0 to p. To get an estimate of the hyperbolic length of γ, we use any partition $0 = t_1 < t_2 < t_3 < \cdots < t_n = 1$ of the unit interval, and estimate the integral in Definition 2.1 by

$$P = \sum_{i=1}^{n-1} \rho(\gamma(t_i)) |\gamma(t_{i+1}) - \gamma(t_i)|.$$

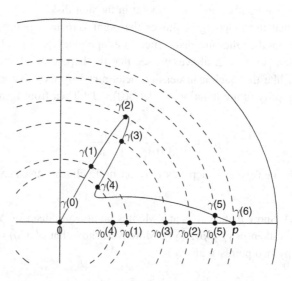

Figure 2.1   Estimating the length of γ

We use the same partition $\{t_i\}$ to compare $\rho(\gamma_0)$ and $\rho(\gamma)$. We project the points $\gamma(t_i)$ radially to the points $\gamma_0^i = |\gamma(t_i)|\frac{p}{|p|}$ on the straight line γ_0. Joining γ_0^i to γ_0^{i+1} by straight line segments, we obtain a path which may move back and forth over itself, but definitely stays on the straight line γ_0. Thus, an estimate for an upper bound for the hyperbolic length of γ_0 is

$$P_0 = \sum_{i=1}^{n-1} \rho(\gamma_0^i) |\gamma_0^{i+1} - \gamma_0^i|.$$

By the invariance of $\rho(t)$ under rotation about the origin,

$$\rho(\gamma_0^i) = \rho(\gamma(t_i)),$$

and, by Euclidean geometry,

$$|\gamma_0^{i+1} - \gamma_0^i| \leq |\gamma(t_{i+1}) - \gamma(t_i)|,$$

so that

$$P_0 \leq P.$$

Since the partition was arbitrary, $\rho(\gamma_0) \leq \rho(\gamma)$. Thus γ_0 has the shortest hyperbolic length among all paths joining 0 to p, and

$$\rho(0, p) = \frac{1}{2} \log \frac{1 + |p|}{1 - |p|} \tag{2.3}$$

is the hyperbolic distance between the origin 0 and any point p in the unit disk.

Suppose now that a and w are any two points in the unit disk. Hyperbolic distance is invariant under Möbius transformations of the unit disk. Therefore, if

$$A(z) = \frac{z - a}{1 - \overline{a}z},$$

then

$$\rho(a, w) = \rho(0, s),$$

where

$$s = A(w) = \frac{w - a}{1 - \overline{a}w}.$$

Using formula (2.3) we obtain the formula for the hyperbolic distance between any two points in the unit disk:

$$\rho(a, w) = \rho(0, s) = \frac{1}{2} \log \frac{1 + |s|}{1 - |s|},$$

$$\rho(a, w) = \frac{1}{2} \log \frac{|1 - \overline{a}w| + |w - a|}{|1 - \overline{a}w| - |w - a|}. \tag{2.4}$$

In analogy with our definition of geodesics for the Euclidean plane we define

Definition 2.2 *If γ is a curve in the hyperbolic plane such that for every triple of points p, r, q on γ with r between p and q we have*

$$\rho(p, q) = \rho(p, r) + \rho(r, q),$$

we call γ a hyperbolic geodesic or a hyperbolic line.

It is clear from the definition that connected pieces of geodesics are also geodesics. As in the Euclidean case, the curve γ may be finite (a path), semi-infinite or infinite, and, if it is geodesic and it is not clear from the context which type of curve it is, we indicate which by respectively calling it a *geodesic segment* or *hyperbolic line segment*, a *geodesic ray* or *hyperbolic ray* or an *infinite geodesic* or *infinite hyperbolic line* or an *extended hyperbolic line*. If it causes no confusion, we omit the adjective hyperbolic.

We can also characterize geodesics in the hyperbolic plane as follows.

Proposition 2.1.1 *A curve γ that realizes the infimum in Definition 2.1 is a geodesic.*

Proof. Suppose first that γ is a curve joining p to q that realizes the infimum and let r be a point on γ. Denote by γ_1 the part of γ that starts at p and ends at r and denote by γ_2 the part of γ that starts at r and ends at q. Then by the definition of the length integral

$$\rho(\gamma) = \rho(\gamma_1) + \rho(\gamma_2)$$

and, since γ is minimal, $\rho(p, q) = \rho(\gamma)$. Then it is easy to see that $\rho(\gamma_1) = \rho(p, r)$ and $\rho(\gamma_2) = \rho(r, q)$. Indeed, if there existed a curve shorter than γ_1 joining p to r, that curve followed by γ_2 would have a shorter length than the curve γ, a contradiction. The same reasoning applies to γ_2 and γ is a geodesic. $\square$

We have showed that there exists at least one geodesic joining zero to any point p in the unit disk, the geodesic $\gamma_0(t) = tp$. Could there be another geodesic, say γ, also joining zero to p? The answer is no as is shown in the next lemma.

Lemma 2.1.1 *For any point $p \neq 0$ in the unit disk, there exists exactly one geodesic $\gamma = \gamma_0(t) = tp$ joining 0 to p.*

Proof. By Proposition 2.1.1 and formula (2.3) the curve γ_0 is geodesic. If γ joining 0 and p is different from γ_0, then there exists a point q with $|q| < |p|$ on γ which is outside γ_0. Let $q_0 = |q|\frac{p}{|p|} \neq q$ be the radial projection of q onto the ray joining the origin and p so that q and q_0 both lie on the circle C centered at the origin with radius $|q|$. If both γ_0 and γ are geodesic,

$$\rho(0, p) = \rho(0, q) + \rho(q, p) = \rho(0, q_0) + \rho(q_0, p).$$

Since $|q| = |q_0|$, $\rho(0, q) = \rho(0, q_0)$ so that

$$\rho(q, p) = \rho(q_0, p). \tag{2.5}$$

We claim this cannot be.

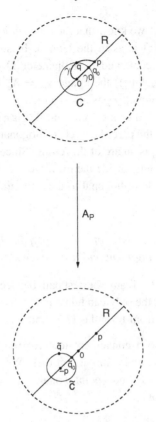

Figure 2.2 The uniqueness of geodesics

Let $A_p(z) = (z - p)/(1 - \bar{p}z)$ be the Möbius transformation taking p to 0. Let $\tilde{q} = A_p(q)$ and $\tilde{q}_0 = A_p(q_0)$ so that $\rho(q, p) = \rho(\tilde{q}, 0)$ and $\rho(q_0, p) = \rho(\tilde{q}_0, 0)$. These distances are realized by the straight lines joining 0 to $\tilde{q}$ and $\tilde{q}_0$ respectively. It follows from formula (2.3) that equation (2.5) holds if and only if $|\tilde{q}| = |\tilde{q}_0|$. Now A_p maps the ray R through 0, q_0 and p to a ray $\tilde{R} = -R$ through $-p$, $\tilde{q}_0$ and 0 and it maps the circle C to a circle $\tilde{C}$ orthogonal to $\tilde{R}$ with hyperbolic center $-p$ and passing through the points $\tilde{q}_0$ and $\tilde{q}$. $\tilde{C}$ therefore separates 0 from its Euclidean center so that $\tilde{q}_0$ is closer to 0 than $\tilde{q}$, contradicting $|\tilde{q}| = |\tilde{q}_0|$. $\square$

From the lemma we obtain the more general statement

Theorem 2.1.1 *For any two points p and q in the unit disk, there exists exactly one geodesic $\gamma = \gamma_{pq}$ joining p to q and it is an arc of a circle orthogonal to the unit disk.*

Proof. By Lemma 2.1.1 we know that there is a unique geodesic γ_0 through 0 and $A_p(q)$ where $A_p(z)$ is as in the lemma. In addition, this geodesic is a straight line segment contained in a diameter D of the disk. It follows that since $\rho(p, q) = \rho(0, A_p(q))$ the image $\gamma_{pq} = A_p^{-1}(\gamma_0)$ is also a geodesic. Moreover, since γ_0 is unique, γ_{pq} is also unique.

Since D is orthogonal to the unit circle, and A_p^{-1} is a Möbius transformation, it follows that $A_p^{-1}(D)$ is the part of a circle orthogonal to the unit circle inside the unit disk and that γ_{pq} is an arc of this circle. Since p and q can be chosen as any pair of distinct points inside the disk, we see that every geodesic is a connected piece of a circle orthogonal to the unit circle. $\square$

An immediate corollary is

Corollary 2.1.1 *Given any pair of points p, q in the unit disk there is a unique shortest path joining them and it is the geodesic γ_{pq}.*

Proof. By Proposition 2.1.1, any shortest path between p and q is a geodesic. Thus, by Theorem 2.1.1, the only candidate is γ_{pq}. That it is the shortest path joining p and q follows from formulas (2.3) and (2.4). $\square$

The cross ratio is also a convenient tool for reformulating the distance between any two points p and q in the unit disk. We begin with the formula for the distance between zero and a point a on the interval $(0, 1)$. The formulas (2.3) and (1.1) tell us that

$$\rho(0, a) = \frac{1}{2} \log cr(-1, a, 1, 0).$$

Observe that -1 and 1 are the intersection points of the hyperbolic line $(-1, 1)$ with the unit circle and this hyperbolic line contains the points 0 and a. Note also that $cr(1, a, -1, 0)cr(-1, a, 1, 0) = 1$. Therefore, the formula that expresses the hyperbolic distance in terms of the cross ratio is

$$\rho(p, q) = \frac{1}{2} |\log cr(x, q, y, p)|, \tag{2.6}$$

where x and y are the endpoints on the unit circle of the geodesic that goes through p and q.

Exercise 2.2 Find the formula for the hyperbolic line that contains the points $\frac{1}{2}$ and $\frac{i}{3}$.

Exercise 2.3 Let γ_1 be the infinite hyperbolic line in the unit disk through the points $-\frac{2}{3}$ and $\frac{2}{3}$, and let γ_2 be the infinite hyperbolic line through the points $\frac{1+i}{3}$ and $\frac{1-i}{3}$. Find the intersection point of the lines γ_1 and γ_2.

Exercise 2.4 Find the set of points in the unit disk Δ which are equidistant from points 0 and $\frac{1}{3}$ with respect to the hyperbolic distance.

Exercise 2.5 Let l be an infinite geodesic in Δ. Show that, for any point $z \in \Delta$, there is a point w on l closest to z where w is the endpoint on l of the geodesic segment through z orthogonal to l; w is called the radial projection from z to l.

Exercise 2.6 The set of all points in the unit disk Δ whose hyperbolic distance from the point $\frac{1+i}{3}$ is equal to 3 is a Euclidean circle. Find the center of that circle.

2.1.2 Hyperbolic triangles

Consider the real axis as a hyperbolic line in Δ and pick a point it, $0 < t < 1$. It is easy to construct many different hyperbolic lines through it that do not intersect the real axis so that the parallel postulate of Euclidean geometry does not hold.

If α, β, γ are hyperbolic lines that intersect in pairs they form a hyperbolic triangle. We can measure the interior angles as the angles between the tangents to the geodesic sides.

Let (b, c) be any two points in the disk that do not lie on the same diameter. Form the Euclidean triangle t with vertices $(0, b, c)$ and the hyperbolic triangle T with the same vertices. The line segments $(0, b)$ and $(0, c)$ are sides of both these triangles. Now compare the Euclidean and hyperbolic lines joining b and c. As the arc of a circle orthogonal to the unit circle, the hyperbolic segment clearly makes a smaller angle with each of the other sides than the Euclidean line does. It follows that the sum of the angles of T is less than the sum of the angles of t and so is less than π.

Two sides of a hyperbolic triangle may meet at the boundary of the hyperbolic plane. As circles orthogonal to the unit circle they are tangent where they meet so the angle between them is zero.

Definition 2.3 *If all three angles of a hyperbolic triangle are zero, it is called* an ideal triangle.

We also have hyperbolic polygons which are domains whose boundary consists of geodesics. We will use these later, in Chapter 6. Note that all the vertices of a polygon may be on the unit circle; in this case, the angles where they meet are zero and the polygon is called an *ideal polygon*.

Exercise 2.7 Let p be any point on the boundary of the unit disk in the upper half plane. Show that any geodesic ray that starts at $\frac{i}{2}$ and terminates at p does not intersect the geodesic line $(-1, 1)$. Find all points p with the property that the infinite geodesic line through p and $\frac{i}{2}$ does not intersect the infinite geodesic $(-1, 1)$.

Exercise 2.8 Let γ_1 and γ_2 be intersecting hyperbolic geodesics with respective endpoints (p_1, q_1) and (p_2, q_2) on the unit circle. Show that

$$cr(p_1, p_2, q_1, q_2) = -\tan^2(\theta/2)$$

where θ is the angle between γ_1 and γ_2 measured along the unit circle from q_1 to q_2. It follows that the lines are orthogonal if and only if the cross ratio is -1.

Exercise 2.9 Let γ_1 and γ_2 be non-intersecting hyperbolic geodesics with respective endpoints (p_1, q_1) and (p_2, q_2) on the unit circle. Show, if u, v are the endpoints of a geodesic δ, then δ is orthogonal to both γ_1 and γ_2 if and only if the cross ratios $cr(p_1, u, q_1, v) = -1$ and $cr(p_2, u, q_2, v) = -1$.

Exercise 2.10 Prove that, if A is a Möbius transformation of Δ that maps z to 0 and w to r, then

$$\cosh(2\rho(0, r)) = \frac{1 + |r|^2}{1 - |r|^2}$$

and thus

$$\sinh^2(\rho(0, r)) = \frac{|r|^2}{1 - |r|^2} = \frac{|w - z|^2}{|1 - \bar{z}w|^2 - |w - z|^2} = \sinh^2(\rho(z, w)).$$

Exercise 2.11 Let T be a hyperbolic triangle with vertices at $(v_a, 0, v_b)$. If $\rho(v_a, v_b) = c$, show that

$$\sinh^2(c) = \frac{|v_b - v_a|^2}{|1 - \bar{v}_b v_a|^2 - |v_b - v_a|^2} = \frac{|v_b - v_a|^2}{(1 - |v_a|^2)(1 - |v_b|^2)}.$$

Hint: Use a Möbius transformation to move T and use Exercise 2.10.

Exercise 2.12 Use Exercise 2.11 and the Euclidean cosine rule to obtain a hyperbolic cosine rule for the triangle T: if $\rho(0, v_a) = b$, $\rho(0, v_b) = a$, $\rho(v_a, v_b) = c$ and the angle at the origin is γ, then

$$\cosh 2c = \cosh 2a \cosh 2b - \sinh 2a \sinh 2b \cos \gamma,$$

Exercise 2.13 Use Exercise 2.12 to show that in the triangle T, if α is the angle at vertex v_a and β is the angle at vertex v_b, then

$$\frac{\sinh 2a}{\sin \alpha} = \frac{\sinh 2b}{\sin \beta} = \frac{\sinh 2c}{\sin \gamma}.$$

Exercise 2.14 Suppose that in triangle T the angles are $\alpha = 0$, $\gamma = \pi/2$ and β so that a is the only finite side. Show that

$$\sinh 2a \tan \beta = 1.$$

Exercise 2.15 Let P be the ideal hyperbolic rectangle in the unit disk with vertices at $(e^{i\theta}, -e^{-i\theta}, -e^{i\theta}, e^{-i\theta})$. Let d be the hyperbolic length of the geodesic between the points where its sides meet the real axis and let a be the hyperbolic length of the geodesic between the points where its sides meet the imaginary axis. Show that

$$\sinh a \sinh d = 1. \tag{2.7}$$

Hint: Use Exercise 2.14.

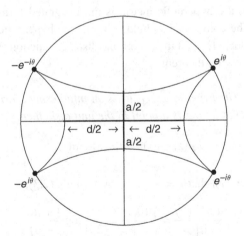

Figure 2.3   Diagram for exercise 2.15

2.2 Properties of the hyperbolic metric in Δ

Now that we have the formula for the hyperbolic metric, we explore some of its basic properties. The first property should, of course, justify calling it a metric.

Theorem 2.2.1 *The unit disk with the hyperbolic metric ρ, (Δ, ρ), is a complete metric space.*

Proof. Definition 2.1 implies that the formula for $\rho(p, q)$ is symmetric and formula (2.4) implies that ρ is non-negative and zero only if $p = q$. The triangle inequality also follows from the definition. This proves (Δ, ρ) is a metric space. Furthermore, in subsection 2.1.1 we showed that except on the unique geodesic joining p and q the triangle inequality is really an inequality.

To prove completeness, suppose that p_n is a Cauchy sequence in (Δ, ρ). Since p_n is a sequence of points in the closed unit disk, there exists a subsequence p_{n_k} which converges in the Euclidean metric to a point p in the closed unit disk. If p is on the unit circle, then, by formula (2.3), $\rho(0, p_{n_k}) \to \infty$. Now fix m; by the triangle inequality $\rho(0, p_{n_k}) \leq \rho(0, p_m) + \rho(p_m, p_{n_k})$, and, by the Cauchy sequence property, $\rho(p_m, p_{n_k}) \to 0$ for all $n_k \geq m$ and $m \to \infty$. It follows that $\rho(0, p_{n_k})$ stays bounded, p is inside the unit disk and formula (2.4) implies $\rho(p, p_{n_k}) \to 0$. Consequently, the Cauchy sequence (p_n) has a convergent subsequence, and thus the whole sequence converges in the hyperbolic metric to the point p. $\square$

By definition, the hyperbolic metric is the integrated form of the hyperbolic density. The converse also holds; that is, the hyperbolic density is the infinitesimal or local form of the hyperbolic distance function. More precisely, we have the following theorem.

Theorem 2.2.2 *The hyperbolic density is an infinitesimal form of the hyperbolic distance. That is, if z is a point in the unit disk, then, for small $t \in \mathbb{C}$,*

$$\rho(z, z+t) = |t|\rho(z) + o(t). \tag{2.8}$$

Proof. Suppose first that $z = 0$. Then, by formula (2.3),

$$\frac{\rho(z, z+t)}{|t|} = \frac{\rho(0, t)}{|t|} = \frac{1}{2|t|} \log \frac{1 + |t|}{1 - |t|},$$

and by L'Hospital's rule

$$\frac{\rho(0, t)}{|t|} \to 1 = \rho(0) \text{ as } t \to 0. \tag{2.9}$$

Suppose now that z is any point in the unit disk and apply the Möbius transformation $A(w) = (w - z)/(1 - \bar{z}w)$. By the invariance of ρ under Möbius transformations proved in Exercise 2.1, we obtain

$$\frac{\rho(z, z+t)}{|t|} = \frac{\rho(0, \frac{t}{1-\bar{z}(z+t)})}{|t|}$$

$$= \frac{\rho(0, \frac{t}{1-\bar{z}(z+t)})}{\left|\frac{t}{1-\bar{z}(z+t)}\right|} \frac{1}{|1-\bar{z}(z+t)|}.$$

Letting $t \to 0$ and using (2.9), we obtain

$$\frac{\rho(z,z+t)}{|t|} \to \frac{1}{|1-\bar{z}(z+0)|} = \rho(z). \quad \square$$

Observe that $|t|$ in formula (2.8) is the Euclidean distance between z and $z + t$. Thus, Theorem 2.2.2 is an infinitesimal comparison between the hyperbolic and Euclidean densities.

Exercise 2.16 Show that (Δ, d), the unit disk with the Euclidean metric, is an incomplete metric space.

Exercise 2.17 Is the function $\rho(z, w)$ continous on $\Delta \times \Delta$?

Definition 2.4 *A domain $D \subset \Delta$ is called* hyperbolically convex *if, for every pair of points $z, w \in D$, the geodesic γ joining z to w is contained in D.*

Exercise 2.18 Show that every hyperbolic disk $D \subset \Delta$ is hyperbolically convex.

2.3 The upper half plane model

An alternative model for hyperbolic geometry is obtained by beginning with the upper half plane $\mathbb{H}$ and using its group of Möbius transformations as isometries.

In Section 1.6 we saw that the group of Möbius transformations of $\mathbb{H}$ is the group consisting of Möbius transformations of the form

$$A(z) = \frac{az+b}{cz+d}, \quad ad-bc = 1, \quad a, b, c, d \in \mathbb{R}.$$

It is easy to compute that

$$\Im A(z) = \frac{\Im z}{|cz+d|^2} \quad \text{and} \quad A'(z) = \left(\frac{1}{cz+d}\right)^2.$$

We could make an analogous derivation for the formula for the hyperbolic metric in this setting. Instead, however, we will use what we have done above to derive the formula, which we denote by $\rho_{\mathbb{H}}(p, q)$, for the hyperbolic distance between any pair of points (p, q) in $\mathbb{H}$. It will follow directly from this discussion that $(\mathbb{H}, \rho_{\mathbb{H}}(p, q))$ is a complete metric space.

For an arbitrary $p \in \mathbb{H}$, set

$$B_p(z) = \frac{z - p}{z - \bar{p}}$$

and note that $B_p(p) = 0$ and $B_p(x) \in \partial \Delta$ for $x \in \mathbb{R}$. If γ is the diameter through 0 and $B_p(q)$ with endpoints $\pm e^{i\theta}$, for some θ, then $B_p^{-1}(\gamma)$ is a semi-circle in $\mathbb{H}$ orthogonal to the real axis with respective endpoints $x, y \in \mathbb{R}$. From equations (2.3) and (2.6) we have

$$\rho(0, B_p(q)) = \frac{1}{2} \log \frac{1 + |B_p(q)|}{1 - |B_p(q)|} = \frac{1}{2} |\log cr(-e^{i\theta}, B_p(q), e^{i\theta}, 0)|. \quad (2.10)$$

Since the cross ratio is invariant under Möbius transformations we have

$$\frac{1}{2} |\log cr(-e^{i\theta}, B_p(q), e^{i\theta}, 0)| = \frac{1}{2} |\log cr(x, q, y, p)|$$

so we define

$$\rho_{\mathbb{H}}(p, q) = \frac{1}{2} |\log cr(x, q, y, p)|. \quad (2.11)$$

Note that for any $p \in \mathbb{H}$ there is always a Möbius transformation $A_p(z)$ preserving $\mathbb{H}$ that sends p to i. The semi-circle orthogonal to the real axis through p and q is mapped to the semi-circle orthogonal to the real axis through i and $A_p(q)$. By the invariance of the cross ratio we have

$$\rho_{\mathbb{H}}(i, A_p(q)) = \rho_{\mathbb{H}}(p, q),$$

That $(\mathbb{H}, \rho_{\mathbb{H}})$ is a complete metric space follows directly from the completeness of (Δ, ρ) by conjugation. That is, map a Cauchy sequence in $\mathbb{H}$ to Δ, find its limit and pull the limit back.

Since the cross ratio depends continuously on its arguments it is clear that $\rho_{\mathbb{H}}(p, q)$ depends continuously on p and q. We can thus compute the infinitesimal form of this distance function. For any small $t \in \mathbb{C}$,

$$\rho_{\mathbb{H}}(p, p+t) = \rho(0, B_p(p+t)) = \rho\left(0, \frac{t}{2i\Im p + t}\right)$$

so that

$$\frac{\rho_{\mathbb{H}}(p, p+t)}{|t|} = \frac{\rho(0, \frac{t}{2i\Im p + t})}{|t|} = \frac{1}{|2i\Im p + t|} \rho\left(0, \frac{t}{2i\Im p + t}\right) \frac{|2i\Im p + t|}{|t|}.$$

Now applying formula (2.9) we have

$$\lim_{t\to 0}\frac{\rho_{\mathbb{H}}(p,p+t)}{|t|}=\frac{1}{2\Im p}\rho(0)=\frac{1}{2\Im p}.$$

We therefore define the density[2] $\rho_{\mathbb{H}}(p)$ by

$$\rho_{\mathbb{H}}(p)=\frac{1}{2\Im p}.$$

The existence and uniqueness of geodesics in this metric space follow again by conjugation to (Δ, ρ). The geodesics in this case are again lines and circles orthogonal to the boundary.

In Section 7.4 we will obtain these formulas in another way.

Exercise 2.19 The triangle T with vertices $(0, 1, \infty)$ is an ideal triangle in the upper half plane. Show that every ideal triangle is congruent to T; that is, show that there is a Möbius transformation that takes any ideal triangle onto T.

Exercise 2.20 Define the hyperbolic area of a region Ω in the upper half plane as $\iint_{\Omega}\rho_H(z)^2dxdy$. Conclude from the previous exercise that all ideal triangles have the same area and evaluate it.

Exercise 2.21 Let T be a hyperbolic triangle in the upper half plane with vertices $0, \infty$ and $1+i$. Evaluate the area of T.

Exercise 2.22 Let T be a hyperbolic triangle in the upper half plane with vertices ∞, i and $1+i$. Evaluate the area of T.

Exercise 2.23 Let T be a hyperbolic triangle in the upper half plane with vertices $i, 1+2i$ and $3i$. Evaluate the area of T.

Exercise 2.24 Let T be a hyperbolic triangle in the upper half plane with vertices i, ∞ and $e^{i\alpha}$. Evaluate the area of T.

Exercise 2.25 Let T be a hyperbolic triangle in the upper half plane with vertices $e^{i\beta}$ and $e^{i\alpha}$ on the unit circle and the third vertex at ∞. Use the previous exercise to evaluate the area of T.

Exercise 2.26 Let T be a hyperbolic triangle in the upper half plane with angles α, β and γ. Use the previous exercises to show that the area of T is equal to $(\pi - (\alpha + \beta + \gamma))/4$.

---

[2] With the normalization $\rho(0) = 2$ we would have $\rho_{\mathbb{H}}(p) = \frac{1}{\Im p}$.

2.4 The geometry of $PSL(2, \mathbb{R})$ and Γ

We have seen that the groups $PSL(2, \mathbb{R})$ and Γ are groups of isometries of the upper half plane $\mathbb{H}$ and unit disk Δ respectively. In this section we classify these isometries.

2.4.1 Hyperbolic transformations

It is obvious from the definition that all elements of $PSL(2, \mathbb{R})$ have real trace. This is also true for all elements of Γ because the trace is invariant under conjugation in $PSL(2, \mathbb{C})$. Let A be the transformation $A(z) = \lambda z$ of $PSL(2, \mathbb{R})$ and assume $\lambda > 1$. By the definition in subsection 1.5.1, A is hyperbolic. The fixed points of A are 0 and ∞ and the imaginary axis $\Im$ is the unique geodesic joining them. We have

Lemma 2.4.1 *Let* $A(z) = \lambda z$, $\lambda > 1$ *and set*

$$l(A) = \inf_{z \in \mathbb{H}} \rho_{\mathbb{H}}(z, A(z)).$$

Then the infimum is achieved for any $z \in \Im$. *Moreover, if* $z \notin \Im$, $\rho_{\mathbb{H}}(z, A(z)) > l(A)$.

Proof. Let z be any point in $\mathbb{H}$. The radial projections of z and λz onto the imaginary axis are, respectively, $|z|i$ and $\lambda|z|i$. By formulas (2.10) and (2.11) we have

$$\rho_{\mathbb{H}}(z, \lambda z) = \rho\left(0, \frac{(\lambda - 1)z}{\lambda z - \bar{z}}\right)$$

$$= \frac{1}{2} \log \frac{1 + |\frac{(\lambda-1)z}{\lambda z - \bar{z}}|}{1 - |\frac{(\lambda-1)z}{\lambda z - \bar{z}}|}$$

$$\geq \frac{1}{2} \log \frac{1 + \frac{(\lambda-1)|z|}{|\lambda z| + |\bar{z}|}}{1 - \frac{(\lambda-1)|z|}{|\lambda z| + |\bar{z}|}}$$

$$= \frac{1}{2} \log \lambda$$

$$= \rho_{\mathbb{H}}(|z|i, \lambda|z|i).$$

If equality holds, then $|\lambda z - \bar{z}| = |\lambda z| + |\bar{z}|$ and the strong triangle inequality for the Euclidean metric implies that 0 lies on the Euclidean segment from λz to $\bar{z}$. That in turn implies that z must be on the imaginary axis. $\square$

From this we have

Proposition 2.4.1 *Let* $A(z)$ *be any hyperbolic transformation in* $PSL(2, \mathbb{R})$ *with fixed points* (x, y) *and let* $L(t)$ *be the geodesic in* $\mathbb{H}$ *joining them. Then* $l(A) = \inf_{z \in \mathbb{H}} \rho_{\mathbb{H}}(z, A(z))$ *is achieved for all* $z \in L(t)$ *but not for any other* $z \in \mathbb{H}$.

Proof. We may assume that $x > y$. Consider the Möbius transformation $B = (z - x)/(z - y) \in PSL(2, \mathbb{R})$. It maps x to 0, y to ∞ and the geodesic $L(t)$ to $\Im$. We can apply Lemma 2.4.1 to BAB^{-1} or its inverse and since B is an isometry of $\mathbb{H}$ the proposition follows. $\square$

Note that $L(t)$ is preserved by A. This leads us to define

Definition 2.5 *For any hyperbolic transformation A, the geodesic joining the fixed points of A is called the* axis *of A and denoted by* Ax_A.

and

Definition 2.6 *For any hyperbolic transformation A, the* translation length *of A is defined as* $l(A)$, *the distance A moves a point on its axis.*

Recall that every non-parabolic element $A \in \mathcal{M}$ can be conjugated to $B(z) = \lambda z$, for some $\lambda \neq 1$. This λ is called the *multiplier of A* and denoted by $\lambda(A)$. If A is hyperbolic, $\lambda > 0$ and, if A is elliptic, $\lambda = e^{2\pi i\theta}$.

Writing B as an element of $PSL(2, \mathbb{C})$ we have

$$B = \pm \begin{pmatrix} \sqrt{\lambda} & 0 \\ 0 & \frac{1}{\sqrt{\lambda}} \end{pmatrix}.$$

Thus, for hyperbolic A,

$$|\operatorname{tr} A| = \left| \sqrt{\lambda(A)} + \frac{1}{\sqrt{\lambda(A)}} \right| \qquad (2.12)$$

and

$$l(A) = l(B) = \frac{1}{2} |\log \lambda(A)| = \rho(i, \lambda i) = \int_1^\lambda \frac{dt}{2t}. \qquad (2.13)$$

Therefore we have

$$|\operatorname{tr} A| = 2 \cosh l(A). \qquad (2.14)$$

We can move this discussion to the disk model. In Section 2.3 we found an element $B_p \in \mathcal{M}$ that takes $\mathbb{R}$ to $\partial\Delta$ and $\mathbb{H}$ to Δ. Any such B_p takes circles orthogonal to $\mathbb{R}$ into circles orthogonal to $\partial\Delta$ and hence geodesics go to geodesics. It takes the fixed points of A to the fixed points of $B_p A B_p^{-1}$ and thus takes Ax_A to $Ax_{B_pAB_p^{-1}}$. The trace and the translation length are preserved under the conjugation.

Exercise 2.27 Show that, if A and B are hyperbolic transformations whose axes are disjoint and not tangent at the boundary, then there is a common orthogonal to these axes.

Exercise 2.28 If A and B are hyperbolic transformations whose axes are disjoint and not tangent at the boundary, prove that the cross ratio (a_1, b_1, a_2, b_2) where a_i, b_i are endpoints of Ax_A and Ax_B respectively is always positive.

Exercise 2.29 Show that, if A and B are hyperbolic transformations whose axes intersect, then the cross ratio of the endpoints of the axes (a_1, b_1, a_2, b_2) is always negative.

2.4.2 Parabolic transformations

We have talked about hyperbolic geodesics and hyperbolic disks inside the hyperbolic plane. There is another kind of region that plays a role in our considerations.

Definition 2.7 *Let p be a point on the boundary of the hyperbolic plane in either the disk or the upper half plane model. Let C_p be a Euclidean circle tangent to the boundary of the hyperbolic plane at p. Then the Euclidean disk H_p that C_p bounds is called a* horoball *at p. Note that, if we use the upper half plane model and let p be the point at infinity, then C_∞ is a horizontal line and H_∞ is the Euclidean half plane above C_∞.*

Let $A(z) = z + a$, $a \in \mathbb{R}$, with single fixed point at infinity. Then every horoball H_∞ is mapped to itself by A. More generally we have

Proposition 2.4.2 *Let $B = (\alpha z + \beta)/(\gamma z + \delta)$ be a parabolic transformation in $PSL(2, R)$ with fixed point $p \in \mathbb{R}$. Then B leaves every horoball H_p invariant.*

Proof. The element $D = -1/(z - p)$ conjugates B to A; that is, $A(z) = DBD^{-1}$. Let H_∞ be a horoball at infinity with boundary $\partial H_\infty = \{z | \Im z = t_0\}$. Note that D^{-1} maps the straight line ∂H_∞ to a circle C_p passing through p and depending on t_0. It also maps the real line to itself, sending ∞ to p. Since $\mathbb{R}$ and ∂H_∞ are tangent at infinity, their images are tangent at p. Therefore C_p is tangent to $\mathbb{R}$ at p and the domain $H_p = D^{-1}(H_\infty)$ is a horoball at p.

Now, since A leaves H_∞ invariant, we conclude that B leaves H_p invariant. □

Again let $l(A) = \inf_{z \in \mathbb{H}} \rho_\mathbb{H}(z, A(z))$. We compute that if $z = x + iy$, with $0 \le x \le a$,

$$l(A) \le \int_0^a \frac{dx}{2y} \to 0 \text{ as } y \to \infty.$$

Similarly, for any parabolic B, conjugating by an isometry we have

$$\inf_{z \in \mathbb{H}} \rho_\mathbb{H}(z, B(z)) = 0. \tag{2.15}$$

Thus, unlike the situation for hyperbolic elements, the infimum is not achieved for any $z \in \mathbb{H}$. Since the trace of a parabolic is plus or minus 2, the formula

$$2 = |\text{tr } B| = 2 \cosh l(B)$$

continues to hold. We say that parabolic elements have zero translation length. The horoballs have the following property.

Proposition 2.4.3 *Let B be a parabolic transformation in $PSL(2, R)$ with fixed point $p \in \mathbb{R}$ and let H_p be a fixed horoball at p. Then there is a positive number K, depending on H_p and B, such that, for all $z \notin H_p$, $\rho_\mathbb{H}(z, B(z)) \ge K$.*

Proof. As above, let $D = -1/(z - p)$ and set $A = DBD^{-1}$. Then $A(z) = z + a$ for some $a \in \mathbb{R}$, depending on B, and $D(H_p) = \{z | \Im z > t_0\}$ for some t_0 depending only on the choice of H_p. For any $z = x + it \notin H_\infty$, we compute, using Exercise 2.30 below, that

$$\rho_\mathbb{H}(D^{-1}(z), BD^{-1}(z)) = \rho_\mathbb{H}(z, z + a) = \rho_\mathbb{H}\left(-\frac{a}{2} + it, \frac{a}{2} + it\right)$$

$$= \log\left(\frac{\sqrt{t^2 + (a/2)^2}}{t} + \frac{a}{2t}\right).$$

Now the last term is a decreasing function of t so that we may take

$$K = \rho_\mathbb{H}\left(-\frac{a}{2} + it_0, \frac{a}{2} + it_0\right) = \log\left(\frac{\sqrt{t_0^2 + (a/2)^2}}{t_0} + \frac{a}{2t_0}\right). \quad □$$

Let us now see what happens if we move our discussion to the disk model Δ. Again, as in Section 2.3, we use the element B_p of $\mathcal{M}$ to map $\mathbb{R}$ to $\partial \Delta$ and the upper half plane to the disk. The boundary $\mathbb{R}$ and the circle ∂H_p in the upper half plane model are tangent at p and are carried to the boundary

$\partial\Delta$ and a circle tangent to $\partial\Delta$ at the image of p; thus horoballs in the upper half plane model are taken to horoballs in the disk model.

Exercise 2.30 Prove that, for $z = x + it$,

$$\rho_{\mathbb{H}}(z, z + a) = \rho_{\mathbb{H}}\left(-\frac{a}{2} + it, \frac{a}{2} + it\right) = \log\left(\frac{\sqrt{t^2 + (a/2)^2}}{t} + \frac{a}{2t}\right).$$

2.4.3 Elliptic transformations

We want to characterize the geometry of elliptic transformations of $PSL(2, \mathbb{R})$ or Γ. For hyperbolic and parabolic transformations, it was easier to start with the upper half plane model. For elliptic transformations, it is easier to start with the disk model.

The elliptic transformation $A(z) = e^{2\pi i\theta}z$ is a Euclidean rotation about the origin and maps Δ to itself. The fixed points of A are 0 and ∞ and these are inverse with respect to $\partial\Delta$. The multiplier is $\lambda = e^{2\pi i\theta}$ and the trace is $\operatorname{tr} A = 2\cos\pi\theta$. Note that, since A has a fixed point in Δ, its translation length is zero.

Consider any map

$$D(z) = \frac{z + a}{1 + \bar{a}z}$$

that sends Δ to itself, 0 to the point a and ∞ to the point $1/\bar{a}$. The conjugate, $B = DAD^{-1}$, has fixed points at a and $1/\bar{a}$ and is again elliptic with trace $2\cos\pi\theta$. Euclidean circles with center at 0 are sent to hyperbolic circles with center at a. As an isometry of the disk, therefore, it is enough to look at the behavior of B with respect to the fixed point a inside Δ. Using the conjugation it is easy to see that hyperbolic disks centered at a are rotated by angle $2\pi\theta$. We call B a *hyperbolic rotation*.

Note that we can write

$$\operatorname{tr} B = \operatorname{tr} A = 2\cos\pi\theta = 2\cosh i\pi\theta = 2\cosh\frac{\log\lambda}{2}.$$

We can thus interpret this rotation as a "translation by an imaginary translation length".

To move this discussion to the upper half plane model, we need to note that the element $B_p^{-1} \in \mathcal{M}$ that sends $\partial\Delta$ to $\mathbb{R}$ takes points inverse with respect to $\partial\Delta$ to points conjugate with respect to $\mathbb{R}$. The map $C(z) = B_p^{-1}BB_p$ is thus a hyperbolic rotation of angle $2\pi\theta$ about every hyperbolic disk in $\mathbb{H}$ whose hyperbolic center is $B_p^{-1}(a)$.

Exercise 2.31 1. Find the Euclidean center of the hyperbolic disk with hyperbolic center i and hyperbolic radius r.
2. Find a formula for all rotations about the points $\pm i$. These are all the rotations about hyperbolic disks in $\mathbb{H}$ (or $\bar{\mathbb{H}}$) with hyperbolic center $\pm i$.

Exercise 2.32 Let A and B be hyperbolic elements such that the axes Ax_A of A and $Ax_{A'}$ of $A' = BAB^{-1}$ are disjoint and not tangent at the boundary. Let $2a = l(A)$ be the translation length of A and let $2d$ be the distance between the axes. Assume further that, if M is the common perpendicular to the axes, L is the perpendicular bisector of M and N is the perpendicular to Ax_A at a point at distance a from the intersection of M and Ax_A, then L and N meet at a point on $\partial \Delta$. That is, the geodesics Ax_A, M, L, N form a quadrilateral with three right angles and one zero angle, and the two finite sides have lengths a, d respectively. Prove that

$$\sinh 2a \sinh 2d = 1. \tag{2.16}$$

Hint: Use Exercise 2.15.

2.4.4 Hyperbolic reflections

Let $R_\Im(z) = -\bar{z}$ denote reflection in the imaginary axis and let z, w be points in $\mathbb{H}$. Then, using the formula (2.11) for distance, it is clear that

$$\rho_\mathbb{H}(z, w) = \rho_\mathbb{H}(R_\Im(z), R_\Im(w)).$$

Thus not only is $R_\Im$ a Euclidean isometry, it is also a hyperbolic isometry but it is not conformal – it reverses orientation. Similarly the reflection $R_\Im(z) = \bar{z}$ is an anti-conformal isometry of the disk model of the hyperbolic plane.

Just as for Euclidean geometry, we can define hyperbolic reflections about any line. These are orientation reversing isometries.

Definition 2.8 Let l be any hyperbolic geodesic. For any point $z \in \mathbb{H}$, there is a point w on l closest to z, the radial projection from z to l (see Exercise 2.5). Define the reflection R_l to be the map that sends a point z to the point $R_l(z)$ on the hyperbolic geodesic through z and w on the opposite side of l such that $\rho_\mathbb{H}(z, w) = \rho_\mathbb{H}(R_l(z), w)$. Note that R_l sends a point on l to itself.

Again, as for Euclidean geometry, any conformal isometry of the hyperbolic plane can be written as a product of two reflections. For example, let $A(z) = \lambda z, \lambda > 1$. Let l_1 be the geodesic in $\mathbb{H}$ with center at the origin and radius $1/\sqrt{\lambda}$ and let l_2 be the geodesic that is the upper half of the unit circle. Then

$$R_{l_1}(z) = \frac{1}{\lambda \bar{z}}, \quad R_{l_2}(z) = \frac{1}{\bar{z}}$$

and the product

$$R_{l_2} R_{l_1} = A(z).$$

Exercise 2.33 Show that every hyperbolic transformation A is the product of reflections through two geodesics orthogonal to Ax_A and intersecting Ax_A at points that are $l(A)/2$ apart.

Exercise 2.34 Show that every elliptic transformation A is the product of reflections through two geodesics passing through the fixed point of A, at angle $\theta/2$, where A is a rotation through angle θ.

Exercise 2.35 Show that every parabolic transformation A is the product of reflections through two geodesics γ_1 and γ_2 tangent at the fixed point p of A where γ_2 lies between γ_1 and $A(\gamma_1)$ and the lengths along any horocycle C_p between its intersection points with γ_1 and γ_2 and between its intersection points with γ_2 and $A(\gamma_1)$ are equal. Hint: Prove it first for $p = \infty$.

3
Holomorphic functions

In this chapter we present a collection of basic theorems from the classical theory of complex variables that play a role in future chapters. We include proofs of those that we make serious use of. For the others, we refer the reader to classical texts, for example [1].

3.1 Basic theorems

The Schwarz lemma is one of the most important tools in complex analysis, and is fundamental in all of the topics we cover in this book. Before stating the Schwarz lemma, however, we recall some basic definitions and theorems.

Definition 3.1 *A differentiable map f from a plane domain Ω to a plane domain X is* holomorphic *if $f_{\bar{z}}(z) = 0$ for all $z \in \Omega$. Note that a constant map is holomorphic.*

As we saw in Section 1.3, this means a holomorphic function has a derivative $f'(z)$ and it is conformal at any point where its derivative does not vanish.

Theorem 3.1.1 *If f is holomorphic on a domain Ω, then, for all n, the derivatives $f^{(n)}$ exist and are holomorphic.*

Theorem 3.1.2 *Every power series $\sum_0^\infty a_n z^n$, $a_n \in \mathbb{C}$, defines a holomorphic function inside its circle of convergence. Moreover, if f is holomorphic on a domain Ω, then it can be expanded in a convergent power series in any disk in Ω.*

If f is holomorphic in a punctured disk, $0 < |z - a| < \delta$, then a is called an *isolated singularity of f*. If f is holomorphic in the neighborhood $|z| > M$ of

53

infinity, f has an isolated singularity at infinity. A holomorphic function with singularities is called *meromorphic*.

Theorem 3.1.3 *Let $a \neq \infty$ be an isolated singularity of f. Then f has a representation as a convergent power series*

$$\sum_{n=-\infty}^{-1} a_n(z-a)^n + \phi(z), \quad a_n \in \mathbb{C},$$

where $\phi(z)$ is holomorphic.

- *If $a_n = 0$ for all $n < 0$, then f has a* removable singularity *at a. That is, it can be extended to a function holomorphic at a.*
- *If $a_n = 0$ for all $n < -N < 0$, $a_{-N} \neq 0$, then f has a* pole of order N *at a. It can be extended to take the value infinity at a and it is locally N to 1 in a neighborhood of a.*
- *If $a_n \neq 0$ for infinitely many n, then f has an* essential singularity *at a and takes on every value, except at most one, infinitely many times. Therefore, there is no natural extension of f to the point a.*

The next theorem characterizes the full group of conformal homeomorphisms of the plane.

Proposition 3.1.1 *The affine maps $f(z) = z_0 + cz$ for $z_0, c(\neq 0) \in \mathbb{C}$, are the only conformal homeomorphisms of the plane.*

Proof. We saw in Section 1.3 that these maps are conformal homeomorphisms of the plane. To see that there are no others, suppose $f(z)$ is a conformal homeomorphism of the plane and expand it in a power series about 0:

$$f(z) = f(0) + a_1 z + a_2 z^2 + \cdots$$

Then $g(z) = f(1/z)$ has an isolated singularity at zero. Since f is one to one, by Theorem 3.1.3, the expansion of $g(z)$ can only contain constant and linear terms and $f(z)$ is affine as claimed. $\square$

Theorem 3.1.3 also implies the converse to Proposition 1.5.1 characterizing the full group of conformal homeomorphisms of the Riemann sphere as the group $\mathcal{M}$ of Möbius transformations.

Proposition 3.1.2 *The only conformal homeomorphisms of the Riemann sphere are Möbius transformations.*

Proof. Suppose that f is a conformal homeomorphism of the sphere. We may assume, without loss of generality, that f fixes the point at infinity since we may post-compose with a Möbius transformation that sends $f(\infty)$ to ∞.

Then a restriction of f to $\mathbb{C}$ is a conformal homeomorphism of the plane and, by Proposition 3.1.1, $f(z) = a_0 + a_1 z$.

Since we post-composed our original f with a Möbius transformation and the composition of an affine map with a Möbius map is again a Möbius map we've proved that f must be Möbius. $\square$

Theorem 3.1.4 *A holomorphic map is either an open map or a constant map.*

Theorem 3.1.5 (Liouville's theorem) *If f is holomorphic and bounded in $\mathbb{C}$ it reduces to a constant.*

Theorem 3.1.6 (Cauchy–Morera theorem) *Let f be defined and continuous in a domain Ω. If f is holomorphic in Ω the integral around any closed curve whose interior is a subset of Ω is zero. Moreover, if the integral around every closed circle in Ω vanishes then f is holomorphic.*

Theorem 3.1.7 (Maximum principle) *A non-constant holomorphic map defined on a domain Ω cannot achieve its maximum in Ω.*

Exercise 3.1 A map f defined on a domain Ω is conformal if and only if it is both holomorphic and one to one.

3.2 The Schwarz lemma

The following lemma is at the heart of our subject and has a very simple elegant proof.

Theorem 3.2.1 (Classical Schwarz lemma) *If f is a holomorphic map from the unit disk Δ to itself such that $f(0) = 0$, then $|f(z)| \leq |z|$ and $|f'(0)| \leq 1$. Equality holds if and only if $f(z) = e^{i\theta} z$ for some θ. (See also Theorem 3.2.2.)*

Proof. Consider the function $g(z) = f(z)/z$ defined in $\Delta \setminus \{0\}$. Since $\lim_{z \to 0} g(z) = f'(0)$, g has a removable singularity at the origin and extends as a continuous and holomorphic map $\tilde{g}(z)$ to all of Δ. Now

$$|\tilde{g}(z)| \leq \frac{1}{r}$$

for all z such that $|z| = r$. By the maximum principle, this inequality holds for all $|z| \leq r$. Taking the limit as $r \to 1$ we conclude that $|\tilde{g}(z)| \leq 1$ for all $z \in \Delta$. Furthermore, since $f'(0) = \lim_{z \to 0} g(z) = \tilde{g}(0)$, we have $|f'(0)| \leq 1$.

Suppose that $|f(z)| = |z|$ for some $z \neq 0$. Then $|\tilde{g}(z)| = 1$ and, by Theorem 3.1.7, $\tilde{g}$ is a constant map. On the other hand, if $|f'(0)| = 1$ then $|\tilde{g}(0)| = 1$ and, again by the maximum principle, $\tilde{g}$ is a constant map. $\square$

As a first application, we prove two properties of the hyperbolic metric that represent the hyperbolic version of the Schwarz lemma.

Theorem 3.2.2 (Schwarz–Pick lemma) *If f is a holomorphic map from the unit disk into itself, then f is both an infinitesimal and a global contraction with respect to the hyperbolic metric on Δ. That is,*

$$\rho(f(t))|f'(t)| \leq \rho(t) \text{ for all } t \in \Delta \tag{3.1}$$

and

$$\rho(f(z), f(w)) \leq \rho(z, w) \text{ for all } z, w \in \Delta. \tag{3.2}$$

In particular, if a holomorphic map f from Δ to itself fixes zero, then $|f(z)| \leq |z|$ and $|f'(0)| \leq 1$.

Proof. Suppose that f is an arbitrary holomorphic function from Δ into Δ. Set

$$g(z) = h \circ f \circ h_0(z),$$

where

$$h(z) = \frac{z - f(t)}{1 - \overline{f(t)}z},$$

and

$$h_0(z) = \frac{z + t}{1 + \overline{t}z}.$$

Then $g(0) = 0$. A simple calculation yields

$$\left| f'(t) \frac{1 - |t|^2}{1 - |f(t)|^2} \right| = |g'(0)| \leq 1,$$

which, by formula (2.2) for the density at a point, translates to

$$|f'(t)|\rho(f(t)) \leq \rho(t).$$

Thus, f is an infinitesimal contraction.

To show that f is a global contraction, suppose that z and w are any two points in the unit disk. By Corollary 2.1.1 there exists a shortest path γ joining z to w. Then the curve $f(\gamma)$ joins points $f(z)$ and $f(w)$ and so, by Definition 2.1,

$$\rho(f(z), f(w)) \leq \rho(f(\gamma)) = \int_{f(\gamma)} \rho(t)|dt| = \int_{\gamma} \rho(f(t))|f'(t)||dt|.$$

Applying (3.1) we obtain

$$\rho(f(z), f(w)) \le \int_\gamma \rho(t)|dt| = \rho(\gamma) = \rho(z, w). \quad \square$$

Theorem 3.2.3 *If we have equality in (3.1) for any t or in (3.2) for any pair of distinct points z and w, then f is a Möbius map from the unit disk onto itself, and we have equality in both (3.1) for all t and (3.2) for all pairs of points z and w in Δ. That is, f is both an infinitesimal isometry,*

$$\rho(f(t))|f'(t)| = \rho(t) \text{ for all } t \in \Delta, \tag{3.3}$$

and a global isometry,

$$\rho(f(z), f(w)) = \rho(z, w) \text{ for all } z, w \in \Delta. \tag{3.4}$$

Furthermore, a map f from Δ into Δ is a hyperbolic isometry if and only if it is a Möbius transformation of the form $f(z) = e^{i\theta}(z - a)/(1 - \bar{a}z)$.

Proof. Suppose first that (3.1) is an equality for some t in Δ. Then the composition $g(z) = h \circ f \circ h_0(z)$ satisfies $|g'(0)| = 1$. The classical Schwarz lemma implies that $g(z) = e^{i\theta}z$ is a hyperbolic isometry of the unit disk, and, since by Exercise 2.1 the same holds for mappings h and h_0, we see that f satisfies (3.4). Setting $w = z + t$ in this formula we obtain

$$\rho(f(z), f(z+t)) = \rho(z, z+t),$$

$$\frac{\rho(f(z), f(z+t))}{|f(z+t) - f(z)|} \frac{|f(z+t) - f(z)|}{|t|} = \frac{\rho(z, z+t)}{|t|}.$$

Letting t tend to zero and applying the comparison of hyperbolic and Euclidean densities in Theorem 2.2.2 we obtain

$$\rho(f(z))|f'(z)| = \rho(z).$$

This proves (3.3).

Now suppose that (3.2) is an equality for some pair of distinct points z, w in Δ. Let

$$h_1(t) = \frac{t - f(z)}{1 - \overline{f(z)}t},$$

and

$$h_2(t) = \frac{z + t}{1 + \bar{z}t}$$

and set

$$\tilde{g}(t) = h_1 \circ f \circ h_2(t).$$

so that $\tilde{g}(0) = 0$. Set $w_0 = (w - z)/(1 - \bar{z}w)$. Since h_1 and h_2 are hyperbolic isometries, we conclude that $\rho(\tilde{g}(w_0), 0) = \rho(w_0, 0)$. Formula (2.3) implies $|\tilde{g}(w_0)| = |w_0|$ and therefore we again conclude that $\tilde{g}(t) = e^{i\theta}t$, and that f is a Möbius transformation as claimed. The formula for $f(z)$ follows from Exercise 1.35. The converse from Exercise 2.1. $\square$

3.3 Normal families

Often, we are concerned not just with a single holomorphic function, but with sequences or families of functions. The following theorems are the standard theorems for dealing with sequences and families of holomorphic functions. We state them here for completeness because we will use them in the following chapters. We refer the reader to standard texts for proofs.

Theorem 3.3.1 (Weierstrass) *Let $\{f_n\}$ be a sequence of holomorphic functions defined on a domain Ω in $\mathbb{C}$ and converging locally uniformly to a function f. Then f is holomorphic. Moreover, the derivatives f_n' converge locally uniformly to $f'(z)$.*

Note that local uniform convergence is equivalent to uniform convergence on compact subsets.

Definition 3.2 *Let $\mathcal{F} = \{f\}$ be a family of holomorphic functions defined on a domain X in $\mathbb{C}$. If every sequence $\{f_i\}$ contains a subsequence $\{f_{i_j}\}$ converging locally uniformly or tending uniformly to infinity, then $\mathcal{F}$ is called a* normal family.

By Theorems 3.1.4 and 3.3.1, the limit function of the convergent subsequence is either an open holomorphic function or a constant which may be infinity. Limit functions of a normal family need not belong to the family.

Theorem 3.3.2 (Hurwitz) *Let f_n be a sequence of holomorphic functions defined on a domain D such that $f_n(z) \neq 0$ for all $z \in D$. If $f_n(z)$ converges locally uniformly to $f(z)$, then either $f(z) \neq 0$ for all $z \in D$, or $f \equiv 0$.*

As an immediate corollary we have

Corollary 3.3.1 *If f_n is a locally uniformly convergent sequence of conformal functions on a domain D then any limit function is either one to one or constant.*

Theorem 3.3.3 *Let $\mathcal{F} = \{f\}$ be a family of holomorphic functions defined on a domain Ω in $\mathbb{C}$ and suppose that on every compact subset $K \subset \Omega$ all functions in the family are uniformly bounded. Then $\mathcal{F} = \{f\}$ is a normal family.*

Theorem 3.3.4 (Montel's theorem) *Let $\mathcal{F} = \{f\}$ be a family of holomorphic functions defined on a domain X in $\mathbb{C}$ and suppose there are at least two points in $\mathbb{C}$ omitted by every function in $\mathcal{F} = \{f\}$. Then $\mathcal{F} = \{f\}$ is a normal family.*

3.4 The Riemann mapping theorem

As we shall see, the Riemann mapping theorem allows us to reduce many problems in complex analysis to problems about functions defined on the unit disk. In this section we present a proof of this theorem. To begin, we need some definitions.

Let D be a plane domain and let $\gamma_i : [0, 1] \to D$, $i = 1, 2$, be two curves in D with the same endpoints; that is, $\gamma_1(0) = \gamma_2(0)$ and $\gamma_1(1) = \gamma_2(1)$.

Definition 3.3 *The curves γ_1 and γ_2 are* homotopic *in D if there exists a continuous map*

$$F : [0, 1] \times [0, 1] \to D$$

such that

$$F(0, t) = \gamma_1(t) \text{ and } F(1, t) = \gamma_2(t),$$

$$F(s, 0) = \gamma_1(0) = \gamma_2(0), F(s, 1) = \gamma_1(1) = \gamma_2(1).$$

This definition allows us to define what it means for a domain not to have "holes" in it.

Definition 3.4 *If D is a domain for which every closed curve $\gamma_1(t)$ is homotopic to its endpoint, that is to the curve $\gamma_0(t) \equiv \gamma_1(0)$, then D is called* simply connected.

We will pursue these ideas further in Section 4.2 where we will be interested in domains that do have holes in them and so are not simply connected.

Definition 3.5 *Two plane domains Ω and X are called* conformally equivalent *if there is a conformal homeomorphism h from Ω onto X.*

From the point of view of conformal equivalence, there are only two different simply connected plane domains. This is the content of the following theorem.

Theorem 3.4.1 (Riemann mapping theorem) *Let X be a simply connected domain in $\mathbb{C}$ such that the complement of X contains at least one point. Then there is a conformal homeomorphism ϕ from X onto Δ. Moreover, if $z_0 \in X$ is chosen, and ϕ is normalized so that $\phi(z_0) = 0$ and $\phi'(z_0) > 0$, then ϕ is unique.*

Note that, because the upper half plane is conformally equivalent to the disk, it follows that any simply connected plane domain that is not the whole plane is conformally equivalent to the upper half plane.

We will apply this theorem to non-simply-connected domains by using the concept of the *universal cover* which comes from topology (see Chapter 4).

We break the proof up into several lemmas. The first gives the uniqueness.

Lemma 3.4.1 *Let X and ϕ be as in the Riemann mapping theorem. Then ϕ is unique.*

Proof. Suppose we have two maps ϕ_1 and ϕ_2 satisfying the conditions in the theorem. Then, since they are homeomorphisms, $f = \phi_2\phi_1^{-1}$ is a conformal homeomorphism of Δ and hence a Möbius transformation. Moreover, since $f(0) = 0$ and $f'(0) > 0$, it follows that $f(z) \equiv z$. $\square$

To show the existence of the map ϕ, we begin by proving that there is a conformal homeomorphism of X into a bounded domain.

Lemma 3.4.2 *Let X be a simply connected domain in $\mathbb{C}$ whose complement contains at least one point and let z_0 be a point of X. Then there is a conformal homeomorphism h of X onto a bounded domain in Δ such that $h(z_0) = 0$.*

Proof. Let $a \in \mathbb{C} \setminus X$. Since X is simply connected, there is no closed curve in X going around a. Therefore, choosing a point $z_0 \in X$ and a branch of $\log(z_0 - a)$ we can define a single valued branch of the logarithm $g(z) = \log(z - a)$ for $z \in X$. Note that $g(z)$ is injective since, if $g(z_1) = g(z_2)$, then

$$z_1 - a = e^{g(z_1)} = e^{g(z_2)} = z_2 - a$$

and $z_1 = z_2$. Pick $z_0 \in X$. Then, because g is open, there is a disk D of radius $2R < \pi$ centered at $g(z_0)$ contained in the image $g(X)$. By the injectivity, $D + 2\pi i$ cannot belong to $g(X)$.

We contract, invert and translate to obtain the required h. That is, set

$$h(z) = \frac{R}{g(z) - (g(z_0) + 2\pi i)} + \frac{R}{2\pi i}. \quad \square$$

From now on we assume that X is a bounded domain inside Δ and that $z_0 = 0$.

Lemma 3.4.3 *Let $\mathcal{F}$ be the family of one to one holomorphic functions f defined on X such that, for all $z \in X$, $|f(z)| < 1$ and $f(0) = 0$. Then $f(X) = \Delta$ if and only if*

$$|f'(0)| = \max_{g \in \mathcal{F}} |g'(0)|. \tag{3.5}$$

Note that by Theorems 3.1.4 and 3.3.3 and Corollary 3.3.1 the supremum is achieved in equation (3.5).

Proof of Lemma. To show the necessity, assume $g \in \mathcal{F}$ and $g(X) = \Delta$. For any other $f \in \mathcal{F}$, set $f = h \circ g$ where h maps Δ into itself and $h(0) = 0$. By the Schwarz lemma, $|h'(0)| \le 1$ or

$$|f'(0)| \le |g'(0)|.$$

To show the sufficiency, assume that $f \in \mathcal{F}$ and that there is some $\alpha \in \Delta$ such that $f(z) \ne \alpha$ for all $z \in X$. We will construct a $g \in \mathcal{F}$ such that $|g'(0)| > |f'(0)|$. For $z \in X$, define a branch of the logarithm by

$$F(z) = \log \frac{f(z) - \alpha}{1 - \bar{\alpha} f(z)}.$$

Since $|(f(z) - \alpha)/(1 - \bar{\alpha}f(z))| < 1$, it follows that $\Re F(z) < 0$.
The Möbius transformation

$$G(w) = \frac{w - F(0)}{w + \overline{F(0)}}$$

maps the left half plane $\Re z < 0$ onto the unit disk. To see this note that the image of the imaginary axis

$$G(it) = -\frac{it - F(0)}{\overline{it - F(0)}}$$

is the unit circle and $G(F(0)) = 0$. Now

$$g(z) = G(F(z)) = \frac{F(z) - F(0)}{F(z) + \overline{F(0)}}$$

62 *Holomorphic functions*

is holomorphic and one to one in X. Moreover, $g(0) = 0$ and $|g(z)| < 1$ so that $g \in \mathcal{F}$.

By a simple computation we have

$$g'(0) = \frac{F'(0)}{F(0) + \overline{F(0)}} \text{ and } F'(0) = \left(\bar{\alpha} - \frac{1}{\alpha}\right) f'(0).$$

Then

$$\frac{|g'(0)|}{|f'(0)|} = \frac{1 - |\alpha|^2}{2|\alpha| \log |\frac{1}{\alpha}|}.$$

To complete the proof we need to verify the inequality

$$\frac{1 - t^2}{t} - 2 \log \frac{1}{t} > 0 \text{ for } 0 < t < 1.$$

Note that the derivative of the left hand side satisfies

$$-\left(\frac{1}{t} - 1\right)^2 < 0$$

and its value at $t = 1$ is 0. Thus for $0 < t < 1$ the left hand side is positive. $\square$

We now complete the proof of Theorem 3.4.1

Proof. We need to prove that there exists a function $f \in \mathcal{F}$ for which $|f'(0)|$ is maximal. To this end, let $\mathcal{A}$ be the subset of functions in $\mathcal{F}$ such that $|f'(0)| \geq 1$. It is not empty since it contains the identity.

By Theorem 3.3.3, since the functions in $\mathcal{A}$ are uniformly bounded, they form a normal family. Given a sequence of functions f_n in $\mathcal{A}$, let f_0 be a limit function of some subsequence. Then, by the Weierstrass theorem, both f_0 and f_0' are holomorphic in X. We claim that $f_0 \in \mathcal{A}$. Clearly, $f_0(0) = 0$, $|f_0(z)| \leq 1$, and, since $|f_n'(0)| \geq 1$, $|f_0'(0)| \geq 1$. Moreover, by the maximum principle, $|f_0(z)| < 1$ for $z \in X$ so that $f_0 \in \mathcal{A}$. It is one to one by Corollary 3.3.1.

It follows that, since the set $\mathcal{A}$ is closed, there is a function f_0 in $\mathcal{A}$ that realizes the maximum in formula (3.5). The function

$$\phi(z) = \frac{|f_0'(0)|}{f_0'(0)} f_0(z)$$

is the normalized Riemann mapping of X onto Δ. $\square$

The Riemann mapping of a domain may or may not extend continuously to the boundary of the domain. A sufficient condition that it extend is

Theorem 3.4.2 *If X is a domain bounded by a Jordan curve then the Riemann map extends as a continuous map of $\bar{X}$ onto $\bar{\Delta}$.*

A full discussion of the Riemann mapping theorem and the proof of this extension theorem may be found in [23], Vol II, or [66].

Exercise 3.2 Prove that if u, v satisfy $\Re u < 0, \Re v < 0$ then $\left|\frac{u-v}{u+v}\right| < 1$.

Exercise 3.3 Find the Riemann mapping for the upper half plane.

Exercise 3.4 Find the Riemann mapping for the strip $\{z|0 < x < 1\}$.

Exercise 3.5 Find the Riemann mapping for the half disk $\{z||z| < 1$ and $x > y\}$.

Exercise 3.6 Find the Riemann mapping for the half strip $\{z|0 < x < 1$ and $y > 1\}$.

Exercise 3.7 Find the Riemann mapping for the infinite wedge domain $\{z|x > y > 0\}$.

3.5 The Schwarz reflection principle

The Schwarz reflection principle gives us a way to extend a holomorphic function outside its domain of definition under certain circumstances. The theorem below is not the most general statement, but will suffice for our needs in this book.

Theorem 3.5.1 (The Schwarz reflection principle) *Suppose D is a domain whose boundary contains a half neighborhood of a segment L of the real axis and let $f: D \to D'$ be a holomorphic map such that the limit as z tends to a point on L is real valued. Set*

$$\overline{D} = \{z|\bar{z} \in D\}.$$

Then f can be extended to L by continuity and it can be extended to the domain $\Omega = D \cup L \cup \overline{D}$ as a holomorphic function F by the rule

$$F(z) = f(z), z \in D \cup L, \qquad F(z) = \overline{f(\bar{z})}, z \in \overline{D}.$$

Thus if $z \in D \cap \overline{D}$ then $f(z) = \overline{f(\bar{z})}$.

Proof. Suppose that we have a holomorphic function g that is defined in the domain Ω and is real valued on the segment L. It is easy to verify that the function $\overline{g(\bar{z})}$ is defined and holomorphic in Ω.

Consider the function $g(z) - \overline{g(\bar{z})}$. Since $z = \bar{z}$ and g is real valued for z on L, the function is zero on L. By Theorem 3.1.2, the function is identically zero in Ω so that $g(z) = \overline{g(\bar{z})}$ for all $z \in \Omega$.

The extended function F is holomorphic in D and $\overline{D}$. We need only prove that it is holomorphic at points of L. To see this, let z_0 be a point on L and let γ bound a disk U such that $z_0 \in U \subset \Omega$. Let γ_1 be the restriction of γ to D and γ_2 the restriction to $\overline{D}$. Denote the subsegment of L contained inside U by l. Then

$$\int_\gamma F dz = \int_{\gamma_1} F dz + \int_l F dz - \int_{\gamma_2} F dz - \int_{-l} F dz = 0$$

and, by the Cauchy–Morera theorem, f is holomorphic at z_0. $\square$

Exercise 3.8 Verify that the function $\overline{f(\bar{z})}$ of Theorem 3.5.1 is holomorphic in $\overline{D}$.

Exercise 3.9 Prove the Schwarz reflection principle for domains bounded by circular arcs. That is prove that, if D is a domain whose boundary contains a half neighborhood of an arc L of the unit circle and if $f : D \to D'$ is a holomorphic map such that the limit as z tends to a point on L lies on the unit circle, then, if

$$\overline{D} = \left\{ z \left| \frac{1}{\bar{z}} \in D \right. \right\},$$

f can be extended to L by continuity and it can be extended to a function F on the domain $\Omega = D \cup L \cup \overline{D}$ as a holomorphic function by the rule

$$F(z) = f(z), z \in D \cup L, \qquad F\left(\frac{1}{\bar{z}}\right) = \frac{1}{\overline{f(z)}}, z \in \overline{D}.$$

Thus, if $z \in D \cap \overline{D}$, then $f(1/\bar{z}) = 1/\overline{f(z)}$.

3.6 Rational maps and Blaschke products

We recall the basic facts about rational maps. A rational map of the Riemann sphere is a map of the form

$$R(z) = \frac{P(z)}{Q(z)}$$

where $P(z)$ and $Q(z)$ are polynomials. We may assume without loss of generality that they have no common factors. The *degree* of $R(z)$ is the maximum of the degrees of P and Q. $R(z)$ takes the value infinity at the zeros of Q; these are the *poles* of R (see Exercise 3.10 below). The value of R at infinity is

$$\lim_{z \to 0} R(1/z)$$

and the derivative at infinity is

$$\lim_{z \to 0} R'(1/z)(-1/z^2).$$

Thus R is a holomorphic map of the Riemann sphere onto itself whose only singularities are poles. This shows that a rational map can be characterized by

Proposition 3.6.1 *A rational map is a holomorphic degree-n map of the Riemann sphere onto itself whose only singularities are poles.*

Exercise 3.10 Show that, at the zeros of Q, $R(z)$ has a Laurent expansion with finitely many negative powers, justifying our statement that these are poles of R.

Exercise 3.11 Show that rational maps

1. are the only holomorphic maps of the Riemann sphere to itself,
2. are conformal, except at the critical points, where the derivative is zero,
3. have the property that each point in the sphere has the same number of pre-images, counted with multiplicity, and this number is the degree of the map.

A special class of rational maps are those that map the unit disk onto itself. By Schwarz reflection in the circle, Exercise 3.9, they map the complement of the disk to itself. These maps are called *Blaschke products*.

Suppose the unit disk is mapped onto itself by an n to 1 holomorphic map. All the zeros of the function must lie inside the disk. Assume they occur at the not necessarily distinct points a_k, $|a_k| < 1$, $k = 1, \ldots, n$. Consider the function

$$A(z) = e^{i\theta} \prod_{k=1}^{n} \frac{z - a_k}{1 - \bar{a}_k z}$$

where $\theta \in [0, 2\pi]$. As we saw in Exercise 1.35, for $|z| < 1$, each factor has modulus less than 1 and if $|z| = 1$ then $|A(z)| = 1$. It follows that $A(z)$ maps the unit disk onto itself and is thus a Blaschke product of degree n.

Note that the poles of the Blaschke product are at the points $1/\bar{a}_k$, $k = 1, \ldots, n$, and these are outside the unit disk. $A(z)$ is conformal at the zeros and poles if and only if they are all distinct; that is, if and only if the zeros and poles are all simple.

Using the following lemma we obtain a stronger statement.

Lemma 3.6.1 *Suppose f is a holomorphic n to 1 map of the disk onto itself. If z_j is a sequence of points in Δ converging to a point on the boundary of the unit disk then $|f(z_j)|$ converges to 1.*

Proof. Suppose for some sequence z_j, $|z_j| \to 1$, we have $f(z_j) \to w$, $|w| < 1$. Let $\zeta_1, \zeta_2, \ldots, \zeta_n$ be the n pre-images of w. Then we can find disjoint neighborhoods N_k of ζ_k on which f is $i(k)$ to 1, and contained in a compact subset K of Δ. Let $V = \bigcap_1^n f(N_k)$; $w \in V$ and, since f is holomorphic, V is open. Then, for every $v \in V$, $\{f^{-1}(v)\} \in K$, and in particular, for j sufficiently large, $z_j \in K$, contradicting $|z_j| \to 1$. $\square$

Theorem 3.6.1 *A holomorphic function is an n to 1 surjective self map of the unit disk if and only if it is a Blaschke product of degree n.*

Proof. Suppose f is a holomorphic n to 1 map of the disk onto itself and let $a_1, \ldots, a_n$ be its zeros counted with multiplicity. Write

$$f(z) = \prod_{k=1}^n \frac{z - a_k}{1 - \overline{a}_k z} g(z)$$

where $g(z)$ is holomorphic and non-zero on Δ. Then $1/g(z)$ is also holomorphic and nonzero on Δ. By the maximum principle, if g is open, either $m = \inf_{z \in \Delta} |g(z)|$ or $M = \inf_{z \in \Delta} |1/g(z)|$ is less than 1. Assume, without loss of generality, that $m < 1$ and let z_j be a sequence such that $\lim_{j \to \infty} |g(z_j)| = m$. By Lemma 3.6.1, it follows that $z_j \to \zeta \in \Delta$ and by continuity it follows that $|g(\zeta)| = m$. Since g is open, either $m = 0$ or m cannot be a minimum and in either case we have a contradiction.

Thus g is a constant k and by Lemma 3.6.1 $|k| = 1$. Thus f extends to the whole sphere and is a Blaschke product as claimed. $\square$

Exercise 3.12 Let $c \neq 0$ be any point in Δ such that $\rho(0, c) < 1$. Set

$$A_a(z) = \frac{z(z - a)}{1 - \overline{a}z}.$$

Let $A_a^{-1}(c) = \{z_1, z_2\}$. Show that $\rho(0, z_1) = \rho(a, z_2)$ and that the points can be ordered so that

$$\rho(0, z_1) = \rho(a, z_2) \to \rho(0, c) \text{ as } |a| \to 1.$$

3.7 Distortion theorems

The condition that a function be holomorphic has geometric implications. If the derivative does not vanish at a point, this is true in a whole neighborhood of the point. Not only is the map conformal at every point in this neighborhood, all the directional derivatives at each point are equal. This means that the shape of the image of a disk about each point isn't too distorted. We make this idea precise below.

If f is holomorphic at p and $f'(p) \neq 0$, a local inverse can be defined in a small disk centered at $f(p)$. As a map from this disk to its image, f is conformal. Another term for this is *univalent*. Pre-composing and post-composing with affine maps, we may assume $p = 0$, $f(0) = 0$ and $f'(0) = 1$.

Here we state three theorems due to Koebe. Proofs can be found in the standard literature, for example, [51].

Theorem 3.7.1 (Koebe$\frac{1}{4}$-theorem) *Let f be univalent in the unit disk Δ and assume $f(0) = 0$ and $f'(0) = 1$. Then $f(\Delta)$ contains a disk of radius at least $\frac{1}{4}$ about the origin.*

Theorem 3.7.2 *Let f be univalent in the unit disk Δ and assume $f(0) = 0$ and $f'(0) = 1$. Then for all z, $|z| < 1$,*

$$\frac{1 - |z|}{(1 + |z|)^3} \leq |f'(z)| \leq \frac{1 + |z|}{(1 - |z|)^3}.$$

Theorem 3.7.3 *Let f be univalent in the unit disk Δ and assume $f(0) = 0$ and $f'(0) = 1$. Then for all z, $|z| < 1$,*

$$\frac{1 - |z|}{(1 + |z|)^2} \leq |f(z)| \leq \frac{1 + |z|}{(1 - |z|)^2}.$$

4

Topology and uniformization

In the construction of Euclidean distance in Section 1.8, we define an atlas for our domain in which every point is contained in a puzzle piece, or chart, that looks like a small disk. The overlaps between the pages of the atlas are given by maps that identify parts of the charts on each page.

We use this idea to define smooth surfaces and in particular Riemann surfaces. We work in some generality but in most of the book our surfaces will be domains that sit inside the standard Euclidean plane.

We will assume that we have a well defined topology on our domains and surfaces. This means that, about each point p in the domain, there is a disk $\{z \mid z - p| < \epsilon\}$ contained in the domain. These disks form a basis for the open sets of the domain and they, in turn, satisfy standard conditions for unions and intersections. (See for example [49].) Unless otherwise stated, our domains are connected. This means that they cannot be expressed as the union of disjoint open sets.

We will need to be able to separate points in our domains. That is, we will assume that, given any two distinct points, we can find disjoint disks, each containing one of the points. A space with this property is called a *Hausdorff space*.

4.1 Surfaces

We are now ready to give precise definitions for our notions.

Definition 4.1 *A topological space is a set X with a collection of subsets O called its* open *sets. Both X and the emptyset belong to O. The union of any number of open sets and the intersection of finitely many open sets are open. The closed subsets are the complements of the open subsets.*

68

Definition 4.2 *A topological space X is called a* Hausdorff *space if, for every pair of distinct points p, q in X, there are disjoint open sets U_p and U_q with $p \in U_p$ and $q \in U_q$. That is, points can be separated by open sets.*

Definition 4.3 *A topological space X is called* connected *if it cannot be written as the union of disjoint open subsets.*

Another characterization of a connected space given in Exercise 4.1 below is that the only non-empty subset that is both open and closed is the whole space.

Definition 4.4 *A subset X of a topological space is* compact *if, for every collection of open sets U_α such that $X \subset \bigcup_\alpha U_\alpha$, there is a finite subcollection $U_1, \ldots, U_n$ such that $X \subset \bigcup_{\alpha=1}^n U_\alpha$. The collections U_α are called* coverings.

Definition 4.5 *Let X be a connected Hausdorff space with an atlas of pairs, or* coordinate charts, *(U_p, ϕ_p), defined for each $p \in X$, where each U_p is an open set containing $p \in X$ called a* neighborhood *of p and $\phi_p : U_p \to \mathbb{C}$ is a homeomorphism from the neighborhood U_p onto an open set in $\mathbb{C}$. Then X is a* smooth surface *if, whenever the neighborhoods U_p and U_q have a non-empty intersection U_{pq}, the* overlap map *$\phi_p \phi_q^{-1} : \phi_q(U_{pq}) \to \phi_p(U_{pq})$, together with all its derivatives, is continuous. It is a* Riemann surface *if the overlap maps are holomorphic. We call $z = \phi_p(x)$ the* local coordinate at p. *It is* orientable *if for each overlap map $f_{pq} = \phi_p \phi_q^{-1}$ the Jacobian $J(f_{pq}) > 0$.*

In this book we will always assume that surfaces are orientable.

Example 1: If X is a plane domain there is a single coordinate chart defined in the whole domain X given by the coordinate $\phi_p(z) = z$ of the plane.

Example 2: If X is the Riemann sphere we need two coordinate charts: $\phi_p(z) = z$ is the local coordinate in the finite plane $X \setminus \{\infty\}$ and $\phi_p(z) = 1/z$ is the local coordinate in $X \setminus \{0\}$. The map $z \mapsto 1/z$ is the holomorphic coordinate change defined on the overlap $X \setminus \{0, \infty\}$.

Definition 4.6 *Given two Riemann surfaces X and Y with coordinate charts (U_p, ϕ_p) and (V_q, ψ_q) respectively, and a map $f : X \to Y$, the map f is* holomorphic *if all the overlap maps $\psi_{f(p)} f \phi_p^{-1}$ are holomorphic.*

The *smooth structure* of a surface is, by definition, the collection of local coordinate charts with smooth overlap maps. Similarly, the *complex structure or Riemann surface structure* of a surface is, by definition, the collection of local coordinate charts with holomorphic overlap maps. If the underlying point sets of the Riemann surfaces X and Y are the same, and if there is a holomorphic homeomorphism f from X onto Y, then the coordinate charts of X and Y determine the same complex structure for the underlying space.

Exercise 4.1 Prove that a topological space is connected if and only if the only non-empty subset that is both open and closed is the whole space.

Exercise 4.2 1. Prove that the map $f(z) = az + b$, $a \neq 0$, $b \in \mathbb{C}$, of the plane to itself preserves the holomorphic structure of the plane.
2. Prove that the map $f(x, y) = (ax + b, cy + d)$, $a \neq 0$, b, $c \neq 0$, $d \in \mathbb{R}$, preserves the smooth structure of the plane, and preserves the holomorphic structure if and only if $a = c$.

Exercise 4.3 Prove that any Möbius transformation preserves the complex structure of the Riemann sphere. Show that these are the only such maps.

Exercise 4.4 Prove that any rational map is holomorphic with respect to the structure from Example 2. Prove that any polynomial map is holomorphic with respect to the structure from Example 1.

4.2 The fundamental group

In our discussions about paths in Euclidean space and the disk we saw that geodesics, that is curves between points for which equality holds in the triangle inequality, also have minimal length. For more general surfaces, this is not necessarily true. For example, consider an annulus, $A = \{z \mid 1 < |z| < R\}$. For two points $p, q \in A$ we can get from p to q by going in a clockwise or counterclockwise direction – and the resulting curves usually have different lengths. These curves are different in an essential way; we formalize this in the discussion below.

Suppose that $\gamma_i : [0, 1] \to X$, $i = 1, 2$, with $\gamma_1(1) = \gamma_2(0)$, are curves on a surface X. We can form the *product* curve $\gamma_1 \gamma_2$ by setting

$$\gamma_1 \gamma_2(t) = \gamma_1(2t), \qquad 0 \leq t \leq 1/2,$$

$$\gamma_1 \gamma_2(t) = \gamma_2(2t - 1), \quad 1/2 \leq t \leq 1.$$

Recall the definition of homotopy between two curves from Section 3.4, Definition 3.3. We repeat it here so that this discussion is self contained.

Definition 4.7 *Two curves* $\gamma_1(t)$ *and* $\gamma_2(t)$, $0 \le t \le 1$, *with the same endpoints are called* homotopic *if there is a continuous map* $F : [0,1] \times [0,1] \to X$ *such that*

$$F(0,t) = \gamma_1(t), \quad F(1,t) = \gamma_2(t) \text{ and}$$

$$F(s,0) = \gamma_1(0) = \gamma_2(0), \quad F(s,1) = \gamma_1(1) = \gamma_2(1).$$

We write $\gamma_1 \sim \gamma_2$. *The map* $F(s,t)$ *is called the homotopy from* γ_1 *to* γ_2.

Note that, in the case of the annulus, this captures the difference between the clockwise and counterclockwise paths. Any homotopy from one to the other would have to contain curves that leave A, go through the hole and come back into A.

Given a point $p \in X$ we can restrict our attention to the family of closed curves with initial point and endpoint p; that is curves such that $\gamma(0) = \gamma(1) = p$. It is straightforward to verify (see Exercise 4.5 below) that the homotopy relation is an equivalence relation on these curves. We denote the equivalence class containing γ by $[\gamma]$. We have

Proposition 4.2.1 *The homotopy classes of closed curves at* p *form a group.*

We call this group the *fundamental group of the surface* X *with basepoint* p and denote it by $\Pi_1(X, p)$.

Proof of Proposition. Denote the constant curve by $\iota_p(t) \equiv p$, $0 \le t \le 1$. Then, for any curve $\iota'_p \in [\iota_p]$ and any closed curve γ at p, $\iota'_p \gamma \sim \gamma \iota'_p \sim \gamma$ (see Exercise 4.6 below). Therefore the class $[\iota_p]$ is the *identity* class of the fundamental group. If $\gamma(t)$ is a closed curve at p, then $\gamma(1-t)$ is its inverse; that is, $\gamma(t)\gamma(1-t) \sim \iota_p$. To see this form the homotopy

$$F(s,t) = \left\{ \begin{array}{ll} \gamma(2t), & 0 \le t \le (1-s)/2, \\ \gamma(1-s), & (1-s)/2 \le t \le (1+s)/2, \\ \gamma(2-2t), & (1+s)/2 \le t \le 1. \end{array} \right\} \qquad (4.1)$$

Further, if $\alpha(t), \beta(t), \gamma(t)$ are three closed curves at p, it is easy to show that, while $(\alpha\beta)\gamma$ and $\alpha(\beta\gamma)$ are not the same curve, one is a reparametrization of the other and they are in the same equivalence class (see Exercise 4.7 below). $\square$

In our example of the annulus A, $\Pi_1(A, p)$ is the infinite cyclic group. A representative for each class is given by $\gamma_n(t) = pe^{2\pi i n t}$.

We will assume, unless we state otherwise, that the spaces X we consider are arcwise connected; that is, given any two points $p, q \in X$ there is always at least one curve $\sigma : [0, 1] \to X$ such that $\sigma(0) = p$ and $\sigma(1) = q$.

Proposition 4.2.2 *Let $\Pi_1(X, p)$ and $\Pi_1(X, q)$ be the fundamental groups of the arcwise connected surface X based at p and q respectively. Then $\Pi_1(X, p)$ and $\Pi_1(X, q)$ are isomorphic.*

Proof. Since X is arcwise connected we can find a curve $\sigma(t)$ on X that joins p to q. To every closed curve γ representing a homotopy class in $\Pi_1(X, q)$, we associate the curve $\gamma_\sigma = \sigma^{-1}\gamma\sigma$. It is straightforward to verify (see Exercise 4.8 below) that the homotopy class of γ_σ depends only on the homotopy class of γ. Moreover, the map $[\gamma] \mapsto [\gamma_\sigma]$ is a homomorphism,

$$[(\alpha\beta)_\sigma] = [\sigma^{-1}\alpha\beta\sigma] = [\sigma^{-1}\alpha\sigma][\sigma^{-1}\beta\sigma].$$

The map is surjective and invertible so it is, in fact, an isomorphism.

Note that the isomorphism actually depends only on the homotopy class of σ: if τ is another curve joining p to q and if $\tau \sim \sigma$, then, from Exercise 4.9 below, $\gamma_\tau \sim \gamma_\sigma$ for all closed curves γ based at p. $\square$

It therefore makes sense to talk about the fundamental group $\Pi_1(X)$ without specifying a basepoint.

An important special case is the following.

Corollary 4.2.1 *The space X is simply connected if and only if $\Pi_1(X)$ is trivial; that is every closed curve is homotopic to its basepoint and $\Pi_1(X) = \{[\iota]\}$.*

We next show that homeomorphic surfaces have the same fundamental group. Precisely,

Theorem 4.2.1 *The fundamental group of an arcwise connected space is a topological invariant.*

Proof. Suppose $h : X \to Y$ is a continuous map of the space X onto the space Y. If $\gamma(t)$ is a curve in X with basepoint p then $\gamma_h(t) = h\gamma(t)$ is a curve in Y with basepoint $h(p)$. Moreover, if γ_1 and γ_2 in X are homotopic by the homotopy $F(t, s)$ then γ_{1h} and γ_{2h} are homotopic by the homotopy $F_h(t, s) = hF(t, s)$. Post-composing the formulas in the definition of product curves shows that the homotopy map $hF(t, s)$ preserves products so that h induces a homomorphism $h_* : \Pi_1(X, p) \to \Pi_1(Y, h(p))$. If h is a homeomorphism then h^{-1} induces the homomorphism h_*^{-1} so that h_* is an isomorphism. $\square$

We will also find the following definition for a less strict form of homotopy useful in Chapter 7.

Definition 4.8 *Suppose γ_0 and γ_1 are closed curves on the surface X. If there is a continuous map*

$$F(s, t) : [0, 1] \times [0, 1] \to X$$

such that $F(0, t) = \gamma_0(t)$, $F(1, t) = \gamma_1(t)$ and $F(s, 0) = F(s, 1)$, we say that γ_0 and γ_1 are freely homotopic. This means that, for each $s_0 \in [0, 1]$, the closed curve formed by composing $F(s, 0)$, $0 \leq s \leq s_0$, $F(s_0, t)$ and $F(s, 1)$, $s_0 \geq s \geq 0$, is homotopic to γ_0.

The set of all curves freely homotopic to a curve γ is the free *homotopy class of γ.*

Note that curves in a free homotopy class don't all share a basepoint.

Exercise 4.5 Prove that the homotopy relation is an equivalence relation on the set of closed curves γ with common basepoint p.

Exercise 4.6 Prove that, if $\iota'_p \in [\iota_p]$ where ι_p is the constant curve $\iota_p(t) \equiv p$ and γ is any closed curve with basepoint p, then $\iota'_p \gamma = \gamma \iota'_p = \gamma$.

Exercise 4.7 Prove that, if $\gamma_2(t)$ is a reparametrization of $\gamma_1(t)$, then $\gamma_1 \sim \gamma_2$.

Exercise 4.8 Let σ be a curve in an arcwise connected space X that joins the point p to the point q and let γ_1, γ_2 be closed curves with basepoint q. Prove that $\sigma^{-1} \gamma_1 \sigma \sim \sigma^{-1} \gamma_2 \sigma$.

Exercise 4.9 Show that, if τ and σ are two curves joining p to q on a surface X and if $\tau \sim \sigma$, then, for all closed curves γ on X based at p, it follows that $\tau \gamma \tau^{-1} \sim \sigma \gamma \sigma^{-1}$.

Exercise 4.10 1. Suppose that $\tilde{\gamma}$ is any curve in $\mathbb{H}$ joining the points $e^{i\theta}$ and $\lambda e^{i\theta}$. Show that $\tilde{\gamma}$ is homotopic to the curve $\hat{\gamma}(v) = (1 + v(\lambda - 1))e^{i\theta}$, $v \in [0, 1]$.
2. Now suppose $\tilde{\gamma}$ is any curve in $\mathbb{H}$ joining the points iy and $1 + iy$. Show that $\tilde{\gamma}$ is homotopic to the curve $\hat{\gamma}(v) = v + iy$, $v \in [0, 1]$.

Exercise 4.11 Let $\Delta^* = \Delta \setminus \{0\}$. Is there a homeomorphism from the annulus A defined above onto Δ^*? Is there a C^∞ homeomorphism? Is there a conformal homeomorphism? Describe $\Pi_1(\Delta^*)$.

Exercise 4.12 Let

$$Smileyface = \Delta \setminus \left(\left\{ 0, \frac{1+i}{2\sqrt{2}}, \frac{-1+i}{2\sqrt{2}} \right\} \cup \left\{ z = \frac{1}{2}e^{i\theta} \,\middle|\, -\frac{3\pi}{4} \le \theta \le -\frac{\pi}{4} \right\} \right).$$

Describe $\Pi_1(Smileyface)$.

4.3 Covering spaces

Next, we need the notion of a covering space.

Definition 4.9 *Given any smooth surface X, a* smooth covering space, *or more simply a* cover, *$(\tilde{X}, \pi)$ of X is a smooth surface, $\tilde{X}$, together with a projection map $\pi : \tilde{X} \to X$, the* covering map, *such that π is surjective and a local homeomorphism. The covering is* holomorphic *if π is holomorphic.*

Note that a holomorphic function f from a domain Ω onto a domain X defines a holomorphic cover (Ω, f) of X if and only if f has no critical points in Ω. For example, $(\mathbb{C} \setminus \{0\}, z^2)$ is a holomorphic cover of $\mathbb{C} \setminus \{0\}$ and $(\mathbb{C}, e^z)$ is a holomorphic cover of $\mathbb{C} \setminus \{0\}$.

To pursue this further we make the definitions

Definition 4.10 *Assume $(\tilde{X}, \pi)$ is a holomorphic covering of X. For $p \in X$ let $\gamma(t)$, $0 \le t \le 1$, be a curve with initial point $\gamma(0) = p$. For $\tilde{p} \in \tilde{X}$ with $\pi(\tilde{p}) = p$, let $\tilde{\gamma}(t)$, $0 \le t < s \le 1$, be a curve in $\tilde{X}$ with initial point $\tilde{p} = \tilde{\gamma}(0)$ and satisfying $\gamma(t) = \pi(\tilde{\gamma}(t))$. Then $\tilde{\gamma}(t)$ is called a* local lift *of γ at $\tilde{p}$.*

Definition 4.11 *Assume $(\tilde{X}, \pi)$ is a holomorphic covering of X. For $p \in X$ let $\gamma(t)$, $0 \le t \le 1$, be a curve with initial point $\gamma(0) = p$ and let $\tilde{\gamma}$ be a local lift to $\tilde{p}$, $\pi(\tilde{p}) = p$. If $s = 1$ and $\lim_{t \to 1} \tilde{\gamma}(t)$ exists then $\tilde{\gamma}$ is called a* lift *of γ at $\tilde{p}$.*

Using the fact that π is a local homeomorphism we can prove (see e.g. [3])

Theorem 4.3.1 *Let $\gamma(t)$ be a curve on a Riemann surface X with initial point p and let $(\tilde{X}, \pi)$ be a cover of X. Suppose $\tilde{p}$ in $\tilde{X}$ satisfies $\pi(\tilde{p}) = p$. Then any lift of $\gamma(t)$ at $\tilde{p}$ is unique.*

Proof. Suppose there are two lifts $\tilde{\gamma}_1(t)$ and $\tilde{\gamma}_2(t)$ of γ to $\tilde{p}$. Let J be the set of t such that $\tilde{\gamma}_1(t) = \tilde{\gamma}_2(t)$. It is closed since X is Hausdorff. Now pick $t_0 \in J$ and let $\tilde{U}$ and U be neighborhoods of $\tilde{\gamma}_1(t_0) = \tilde{\gamma}_2(t_0)$ and $\gamma(t_0)$ respectively such that $\pi_{\tilde{U}}$ is a homeomorphism onto U. Thus for all $t \in \tilde{\gamma}_1^{-1}(\tilde{U}) \cap \tilde{\gamma}_2^{-1}(\tilde{U})$ we have $\tilde{\gamma}_1(t) = \tilde{\gamma}_2(t)$ and J is relatively open in $[0, 1]$. Now because J contains 0 it is non-empty and therefore, by Exercise 4.1, it is equal to $[0, 1]$. □

It is not always true that every curve has a lift. This depends on the covering. We therefore define

Definition 4.12 *A covering $(\tilde{X}, \pi)$ of X is called a* regular covering *if, for all $p \in X$, every curve $\gamma(t)$, $\gamma(0) = p$, has a lift to each $\tilde{p} \in \tilde{X}$ with $\pi(\tilde{p}) = p$.*

To understand this concept better, let us look at some non-regular coverings.

The covering $(\mathbb{C} \setminus \{0\}, z^2)$ of $\mathbb{C} \setminus \{0\}$ is regular. Let us remove some point in the cover, say 1, and let $(\tilde{X}, \pi) = (\mathbb{C} \setminus \{0, 1\}, z^2)$. Now let $\gamma(t)$ be the straight segment that starts at $p = \frac{1}{4}$ and ends at 1. Then we can lift γ to a curve $\tilde{\gamma}$ that starts at $\tilde{p} = -\frac{1}{2}$ and ends at -1 but if we form the local lift $\tilde{\gamma}(t)$ with initial point $\tilde{\gamma}(0) = \tilde{p} = \frac{1}{2}$, and satisfying $\pi(\tilde{\gamma}(t)) = \gamma(t)$, we see that $\lim_{t \to 1} \tilde{\gamma}(t)$ is not defined.

Of course this example is artificial. Note, however, that the problem is that the point 1 in X is the only point with only one pre-image.

For the regular covering of $\mathbb{C} \setminus \{0\}$ given by $(\mathbb{C}, e^z)$ there are unbounded curves in $\mathbb{C}$, for example $\tilde{\gamma}(t) = -t, t > 1$, whose image $\gamma(t) = e^{-t}$ is bounded. We use this idea to create a non-regular covering.

The map $f(z) = ze^z$ is holomorphic on $\mathbb{C}$ and has an essential singularity at infinity. It has a critical point at $z = -1$ and critical value $f(-1) = -e^{-1}$. If we let $S = \{z \in \mathbb{C} \mid f(z) = -e^{-1}\}$ and set $\tilde{X} = \mathbb{C} \setminus S$, $(\tilde{X}, f)$ is a covering of $\mathbb{C} \setminus \{-e^{-1}\}$. Let us see that it is not a regular covering. See Figure 4.1.

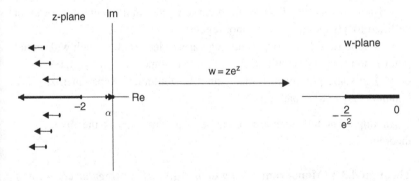

Figure 4.1 Non-regular covering – the arrows indicate the direction of the pre-images of $\gamma(t)$

Restricted to the real axis, $f(x)$ has a minimum at -1 and every point $y \in (-e^{-1}, 0)$ has two real pre-images. Choose some y_0 in that interval and let $x_1 < -1$, $-1 < x_2 < 0$ be its pre-images. Let $\gamma(t) = y_0(1 - t)$ be the straight segment from y_0 to 0. The local lift $\tilde{\gamma}(t) : [0, 1) \to \tilde{X}$ with $\tilde{\gamma}(0) = x_2$ satisfies $\lim_{t \to 1} \tilde{\gamma}(t) = 0$ and so is a lift but the local lift $\tilde{\gamma}(t) : [0, 1) \to \tilde{X}$ with $\tilde{\gamma}(0) = x_1$ satisfies $\lim_{t \to 1} \tilde{\gamma}(t) = -\infty$ and thus cannot be continued to a lift.

Because $f(z)$ is entire, the point y has infinitely many other pre-images. The local lifts of γ to all of these are unbounded and cannot be continued. The only lift that can be continued is the lift to x_2. The point here is that zero has only one pre-image for this function whereas all other points have infinitely many pre-images. See Exercises 4.17, 4.18 below for proofs.

These examples lead us to define

Definition 4.13 *Let* $(\tilde{X}, \pi)$ *be a covering space of X and let* $\gamma(t)$, $0 \leq t \leq 1$, *be a path with initial point* $p = \gamma(0)$ *in X. If there is no lift of* γ *with initial point* $\tilde{p}$ *but there is a local lift* $\tilde{\gamma}(t)$ *with* $0 \leq t < 1$ *then* $\tilde{\gamma}(t)$ *is called an* asymptotic path *for* π *and* $\lim_{t \to 1} \gamma(t)$ *is called an* asymptotic value.

Note that an asymptotic path must leave every compact set in $\tilde{X}$ – otherwise, by the continuity of π it could be continued to a lift.

Regular coverings can be characterized in several different ways. More details can be found in, for example, [3], Chapter I.

Theorem 4.3.2 *A regular covering* is a covering for which there are no asymptotic paths.

Proof. Suppose there is an asymptotic path $\tilde{\gamma}(t)$ in the covering space $(\tilde{X}, \pi)$ and let $\gamma(t)$ be its projection to X with asymptotic value $\lim_{t \to 1} \pi(\tilde{\gamma}(t)) = q$. Then by Theorem 4.3.1 $\gamma(t)$ is a curve with terminal point q that cannot be lifted to $\tilde{\gamma}(0)$ so the cover is not regular.

Now suppose the covering is not regular and let $\gamma(t)$ be a path with initial point p and terminal point q such that there is some local lift $\tilde{\gamma}(t), 0 \leq t < s$, of $\gamma(t)$ to some pre-image $\tilde{p}$ that cannot be continued. Reparametrizing we obtain an asymptotic path. $\square$

An important tool in constructing regular coverings is the monodromy theorem.

Theorem 4.3.3 (Monodromy) *Suppose that* $(\tilde{X}, \pi)$ *is a regular covering of X and let* $\gamma_i(t)$, $i = 1, 2$, *be two curves on X joining p to q such that* $\gamma_1 \sim \gamma_2$. *If* $\tilde{\gamma}_i(t)$, $i = 1, 2$, *are respective lifts with the same initial point* $\tilde{p} \in \tilde{X}$, *then*

they are both well defined and have the same terminal point $\tilde{q}$. Moreover,
$\tilde{\gamma}_1 \sim \tilde{\gamma}_2$ *in* $\tilde{X}$.

Proof. Because the covering is regular, the lifts $\tilde{\gamma}_1$ and $\tilde{\gamma}_2$ to $\tilde{p}$ are well defined connected components of $\pi^{-1}(\gamma_1)$ and $\pi^{-1}(\gamma_2)$ that can be continued to $t = 1$ and so are compact. For any point $\tilde{x}$ on $\tilde{\gamma}_1$, there is a neighborhood $U_{\tilde{x}}$ such that f restricted to $U_{\tilde{x}}$ is a homeomorphism. Since $\tilde{\gamma}_1$ is compact, it has a finite subcover by subsets $\{U_{\tilde{x}_i}\}$.

To show that $\tilde{\gamma}_1$ and $\tilde{\gamma}_2$ are homotopic, let $F(s, t)$ be a homotopy between γ_1 and γ_2 as defined by Definition 4.7. Using the finite subcover of $\tilde{\gamma}_1$, for all sufficiently small s, there exists a homotopy $\tilde{F}(s, t)$ with $\pi\tilde{F} = F$, $\tilde{F}(s, 0) = \tilde{p}$ and $\tilde{F}(s, 1) = \tilde{\gamma}_1(1)$. Moreover, choosing s_0 small enough and applying the above argument to curves $\gamma_{s_0}(t) = F(s_0, t)$ shows that the set of s for which the homotopy $\tilde{F}$ is defined is open. To show that this set is also closed, suppose $\tilde{F}$ is well defined for all s in $[0, s_1)$. Applying Theorem 4.3.1 and the above argument to the curve γ_{s_1}, we conclude that the homotopy can be extended so that, for all s in some neighborhood of s_1, $\tilde{\gamma}_s(1) = \tilde{\gamma}_1(1)$ and the curves $\tilde{\gamma}_s$ are all homotopic. Therefore, the set of s for which the homotopy $\tilde{F}$ can be extended is the whole interval and $\tilde{\gamma}_1$ and $\tilde{\gamma}_2$ are homotopic. $\square$

An immediate corollary of the monodromy theorem is

Corollary 4.3.1 *If* π *is a regular covering from* $\tilde{X}$ *onto* X*, and* p *is any point in* $\tilde{X}$*, then* $\Pi_1(\tilde{X}, p)$ *is a subgroup of* $\Pi_1(X, \pi(p))$*.*

In particular we have

Corollary 4.3.2 *If* $(\tilde{X}, \pi)$ *is a regular covering of a simply connected surface* X *then* π *is a homeomorphism.*

Definition 4.14 *If the regular covering space* $\tilde{X}$ *of a space* X *is simply connected it is called the* universal covering space *of* X*.*

Exercise 4.13 Prove Corollary 4.3.1.

Exercise 4.14 Prove Corollary 4.3.2.

Exercise 4.15 Find a univeral covering π from the unit disk Δ onto the punctured unit disk $X = \Delta \setminus \{0\}$ and verify Corollary 4.3.1 in this case.

Exercise 4.16 Find a univeral covering π from the unit disk onto Δ onto the annulus $A = \{z | 1 < |z| < 3\}$ and verify Corollary 4.3.1 in this case.

Exercise 4.17 Prove that the function $f(z) = ze^z$ takes every value in $\mathbb{C}$. Hint: 1. Show that any omitted value would have to be isolated. 2. Show that any curve whose image ends at an omitted value must be unbounded. 3. Show that all unbounded curves whose image ends at a finite point project to a curve landing at zero.

Exercise 4.18 Prove that, for the function $f(z) = ze^z$, zero has only one pre-image. Prove that every other point has infinitely many pre-images.

Exercise 4.19 Let $f(z) = \int_0^z e^{-t^2} dt$. Show that $f(z)$ is an entire function that has no omitted values. Show that it has two finite asymptotic values, $\pm\sqrt{\pi}/2$.

4.4 Construction of the universal covering space

Theorem 4.4.1 (Existence of a universal cover) *Given any smooth surface X, it admits a universal covering.*

A full discussion of the proof of this theorem is beyond the scope of this book. Instead we give a sketch of the proof. The reader is referred to the standard literature, e.g. [41], [3] and [2], for details.

Proof. [Sketch] We begin by assuming that X is a compact surface of finite genus $g > 0$. The intuitive idea of the construction is the following. We assume that we have a collection $\Gamma = \{\gamma_i\}$, $i = 1, 2, \ldots, n$, of simple closed curves γ_i on X based at a point q such that each γ_i is a representative of a distinct homotopy class and none of the γ_i divide the surface into two pieces. We cut X along each curve in Γ and label each side of the cut $\gamma_i^{\pm}$ according to whether X lies to the left or right as we traverse γ_i. Because none of the curves divide the surface, the result of cutting X yields a connected surface X_0 with boundary. Cutting the surface along each γ_i decreases the genus and we assume there are enough curves that the genus of X_0 is zero – that is, X_0 is simply connected.

The universal cover is formed by taking infinitely many copies $\hat{X}_i$ of $\hat{X}_0$.[1] For $\hat{X}_1$, choose some boundary curve, say γ_1^+, and attach $\hat{X}_1$ to $\hat{X}_0$ by gluing the curve γ_1^+ of $\hat{X}_1$ to the curve γ_1^- of $\hat{X}_0$. Do the same for all the boundary curves. We obtain a bigger surface with new boundary curves on the attached $\hat{X}_i$. Again, for each of the boundary curves $\gamma_i^{\pm}$ of the new surface, we attach

---

[1] See also Section 6.5.

another $\hat{X}_k$ along $\gamma_i^{\mp}$. Since each time we add another copy there are new boundary curves, we can continue the process indefinitely.

Choose a basepoint q_0 on the original cut-up surface X_0. There is a copy q_k of q_0 in each of the infinitely many $\hat{X}_k$ used to form $\tilde{X}$. Any open arc σ_k from q_0 to q_k projects to a closed curve on X and, since $\tilde{X}$ is simply connected, the projection defines an element of $\Pi_1(X)$. All the elements of the fundamental group are obtained this way.

We set $\tilde{X} = \bigcup \hat{X}_i$. Because a point in $\tilde{X}$ lies inside or on the boundary of some $\hat{X}_i$, we define the covering projection π to identify it with the corresponding point in X. Except for points on the copies of γ_i, and vertices where the γ_i meet, the local charts are defined as charts on $\hat{X}$. At the points on the γ_i, the charts are defined by the gluing. Similarly, a neighborhood of a vertex in $\tilde{X}$ projects to a point on X and the chart is defined by the pullback. Since there are only finitely many boundary curves and vertices for each $\hat{X}_i$ the limiting object $\tilde{X}$ is a surface.

If X is smooth, this construction defines a smooth structure on $\tilde{X}$ in the obvious way. The local coordinate $z = \Phi_{\tilde{p}}(\tilde{p})$ where $\pi(\tilde{p}) = p$ and $z = \Phi_{\tilde{p}}(\tilde{p}) = \phi_p \pi(\tilde{p})$ is defined on the component of $\pi^{-1}(U_p)$ containing $\tilde{p}$. The overlap maps $\Phi_p \Phi_q^{-1} = \phi_p \pi \pi^{-1} \phi_q^{-1}$ are obviously smooth at interior points of the copies of $\hat{X}$; at boundary points and vertices one has to use the gluing to check the overlap maps.

If the surface X has finitely many boundary components, the above construction can be modified to include curves from the basepoint to the boundary components. Such a surface is called a surface of "finite topological type".

If the surface X has infinite genus, or infinitely many boundary components, more has to be done. It is necessary to find an "exhaustion" of X by surfaces X_i of finite topological type. That is, X has to be shown to be a limit of the sequence X_i, $X_i \subset X_{i+1}$, in an appropriate sense. The universal cover of X is then a limit of universal covers of the X_i. The fundamental groups inject $i : \Pi_1(X_i) \to \Pi_1(X_{i+1})$ and the fundamental group $\Pi_1(X) = \bigcup_i \Pi_1(X_i)$. $\square$

If X is a Riemann surface, the construction defines the covering as a holomorphic covering since the overlap maps are holomorphic. In fact, choose $\tilde{p} \in \tilde{X}$ and choose the local coordinate so that $\Phi_{\tilde{p}}(\tilde{p}) = 0$.

Exercise 4.20 Show that one can construct a simply connected plane domain with $4g$ boundary curves such that, if they are properly identified in pairs, one obtains a surface X of genus g. Show that the $2g$ identified curves on X generate $\Pi_1(X)$.

4.5 The universal covering group

We need to know the relationship between the pre-images in the covering space of a given point in X under the covering map π. The next theorem gives us that relationship when the covering is the universal covering. Although the theorem can be stated more generally, this is the version that we will need.

Theorem 4.5.1 *If π is a universal covering map from $\tilde{X}$ onto a surface X, and if t and s are two points in $\tilde{X}$ with the same image under π, that is $\pi(t) = \pi(s)$, then there exists a map f from $\tilde{X}$ onto $\tilde{X}$, such that $f(t) = s$ and $\pi(f(\tilde{p})) = \pi(\tilde{p})$ for all $\tilde{p}$ in $\tilde{X}$. Moreover, f is a homeomorphism.*

Proof. We construct the map f as follows. Let $\tilde{p}$ be a point in $\tilde{X}$. Take any path γ that starts at t and ends at $\tilde{p}$. Now project γ down to X, and then lift its projection $\pi(\gamma)$ to a new curve $\tilde{\gamma}$ which starts at the point s. Let $f(\tilde{p})$ be the endpoint of $\tilde{\gamma}$. The map f is well defined because $\tilde{X}$ is simply connected. Indeed, by the monodromy theorem, any two curves in $\tilde{X}$ that start at t and end at $\tilde{p}$ are homotopic, as are their images under π, as well as the lifts of these images to s. Obviously $f(t) = s$ and $\pi(f(z)) = \pi(z)$ for all z in $\tilde{X}$. Since the universal covering map π is locally univalent, we conclude, by taking local inverses of π, that f is open. To see that f is invertible, and hence a homeomorphism, simply construct the inverse of f using the construction for the map f and switching the points t and s. $\square$

An immediate corollary of the construction of the map f in the proof is

Corollary 4.5.1 *A covering map is either fixed point free or the identity.*

This theorem leads us to define

Definition 4.15 *The set S of homeomorphisms of $\tilde{X}$ in the theorem that identify pre-images of the same point in X forms a group G. If π is a universal covering map, and if g is a homeomorphism of $\tilde{X}$, then $g\pi$ is also a universal covering map. Applying the theorem to $g\pi$ we obtain the group $G_g = g^{-1}Gg$ of homeomorphisms of $\tilde{X}$. We therefore define the universal covering group of $\tilde{X}$ as the conjugacy class of G in the group of all homeomorphisms of $\tilde{X}$.*

We have

Theorem 4.5.2 *The universal covering group is isomorphic to $\Pi_1(X, p)$.*

Proof. The isomorphism is constructed exactly as in the proof above with $\tilde{p}$ chosen so that $p = \pi(\tilde{p})$. Note that by the monodromy theorem any homotopically non-trivial closed curve in X must lift to an open curve in

$\tilde{X}$. By Theorem 4.5.1, the covering group is a subgroup of the group of all homeomorphisms of $\tilde{X}$. □

As we saw above, if X is a Riemann surface, then so is $\tilde{X}$. Moreover, by the definition of f in Theorem 4.5.1, it is a conformal homeomorphism of $\tilde{X}$.

4.6 The uniformization theorem

Recall that from the Riemann mapping theorem, Theorem 3.4.1, we obtain a dichotomy for simply connected plane domains: the plane itself and all other simply connected domains. We will apply this theorem to non-simply-connected domains by using the concept of the universal covering surface.

Theorem 4.6.1 (Uniformization theorem) *The universal covering space $\tilde{X}$ of an arbitrary Riemann surface X is homeomorphic, by a conformal map Φ, to either the Riemann sphere, the complex plane or the unit disk and $\Pi_1(X)$ has a representation as a group of conformal homeomorphisms of $\Phi(\tilde{X})$.*

Proof. The universal covering surface $\tilde{X}$ is a simply connected Riemann surface. If we could realize it as a subset of the Riemann sphere by a conformal map $\Psi : \tilde{X} \to \hat{C}$, we would have a trichotomy: the image $\Psi(\tilde{X})$ would either be the full sphere $\hat{C}$ or a simply connected proper subset; in the latter case, $\Psi(\tilde{X})$ would either be the full plane C or have at least two boundary points and we could apply the Riemann mapping theorem to deduce that $\Psi(\tilde{X})$ is conformally equivalent to Δ. The crux of the proof is to construct this mapping. The construction, however, is beyond the scope of this book and we refer the reader to the relevant literature, for example, [2].

Assume we have such an embedding Ψ and if $\Psi(\tilde{X})$ has at least two boundary points let Φ be the Riemann map of $\Psi(\tilde{X})$. Since the fundamental group acts as a group of homeomorphisms of $\tilde{X}$, the map Ψ would induce an isomorphism $\Psi_* : \Pi_1(X) \to G$, where G is a group of conformal homeomorphisms of either $\hat{C}$, C or $\Phi\Psi(\tilde{X})$.

The trichotomy gives the following.

- $\Psi(\tilde{X}) = \hat{C}$ and so is compact.

  Proposition 3.1.2 implies all elements of the covering group are Möbius transformations and so have a fixed point. Now Corollary 4.5.1 implies the only covering map is the identity and $\Pi_1(X) \equiv \{Id\}$ so that X is conformally equivalent to $\hat{C}$.

- If $\Psi(\tilde{X}) = \mathbb{C}$.

  Proposition 3.1.1 implies the covering transformations are affine maps and Corollary 4.5.1 implies any non-trivial covering transformation must be parabolic with fixed point infinity. Thus, in this case, either X is conformally equivalent to $\mathbb{C}$ and $\Pi_1(X) \equiv \{Id\}$, or $G = \Psi_*(\Pi_1(X))$ is a non-trivial subgroup of the complex affine transformations.

- If $\Psi(\tilde{X})$ is a proper subset of $\mathbb{C}$, then there is a conformal Φ such that $\Phi\Psi(\tilde{X}) = \Delta$.

  In this case the covering transformations belong to Γ, the group of Möbius transformations preserving Δ, and $G = (\Phi\Psi)_*(\Pi_1(X))$ is a subgroup of Γ. In this case X is called a *hyperbolic* surface. $\square$

From now on, when we talk about the universal cover $\tilde{X}$ of a Riemann surface X, we will always mean its realization as the sphere, the plane or the disk. The universal covering group then always has a representation as a group of Möbius transformations.

If $\tilde{X}$ is the complex plane we can explicitly write down the universal covering projections. In the simplest case, other than the trivial group, the covering group is the group of translations:

$$\pi(s) = \pi(t) \text{ whenever } s = t + 2\pi n i \text{ for some } n \in \mathbb{Z}.$$

The projection map from the universal cover $\tilde{X} = \mathbb{C}$ is $\pi(z) = e^z + a$ and X is the punctured plane $\mathbb{C} \setminus \{a\}$.

In the next case the universal covering group is the lattice group $\{m + n\tau \mid m, n \in \mathbb{Z}, \Im\tau > 0\}$ and the surface X is a torus. The torus is identified with the quotient of the plane by the group; that is, $\pi(s) = \pi(t)$ if, for some point $m + n\tau$ in the lattice, $s = t + m + n\tau$. We will discuss this further in Section 5.2. The projection map can be written explicitly in terms of elliptic functions. A brief introduction to elliptic functions is given in the appendix.

When $\tilde{X}$ is the unit disk, the covering group is much more interesting. We will discuss these covering groups in detail in the next two chapters. The projection map π is not at all obvious and in fact is almost always hard to find explicitly.

In the latter part of the book, we will focus almost entirely on the situation where X is hyperbolic and we will almost always identify $\tilde{X}$ with the unit disk Δ.

5

Discontinuous groups

5.1 Discontinuous subgroups of $\mathcal{M}$

Let X be a hyperbolic Riemann surface and suppose $\pi : \Delta \to X$ is a universal covering projection. Let p be a point on X and let s and t be points in Δ such that $\pi(s) = \pi(t) = p$. Since π is a local homeomorphism, we may find a small neighborhood V of p such that $\pi^{-1}(V)$ contains disjoint connected components V_s and V_t, containing s and t respectively.

This property motivates us to make the following definition.

Definition 5.1 *A subgroup G of $PSL(2, \mathbb{C})$ or $\mathcal{M}$ is said to act* discontinuously *at $z \in \hat{\mathbb{C}}$ if there is a neighborhood U of z such that $U \cap A(U) = \emptyset$ for all but finitely many $A \in G$.*

The set of points $\Omega(G) \subset \hat{\mathbb{C}}$ at which G acts discontinuously is called the *regular set* of G and the complement $\Lambda(G)$ of the regular set in $\hat{\mathbb{C}}$ is called the *limit set* of G. We often write Ω or Λ if it doesn't cause confusion. The regular set of G is open by definition, but it may be empty. If it is not empty we say that G is a *discontinuous group*.

Discontinuity is invariant under conjugation.

Proposition 5.1.1 *Suppose $G \subset SL(2, \mathbb{C})$ and let B be any element of $SL(2, \mathbb{C})$. Then G and the group G' consisting of all elements of the form $C = BAB^{-1}$ for all $A \in G$ are either both discontinuous or both not discontinuous.*

Proof. It is easy to check from the definition that $\Omega(G') = B(\Omega(G))$. $\square$

It follows directly from the definition that Ω is invariant under G, and therefore that Λ is also. Since Ω is open, Λ is always closed.

It is easy to see the following.

Proposition 5.1.2 *If G is discontinuous then it is discrete.*

Proof. Assume the contrary and let z be any point in $\mathbb{C}$. Let A_n be a distinct sequence of elements of G converging to the identity and let U be a neighborhood of z containing a disk of radius ϵ. Since $A_n \to Id$, we can find an integer N, depending on ϵ, such that $|A_n(z) - z| < \epsilon$ for all $n > N$. Thus $A_n(U) \cap U \neq \emptyset$ for $n > N$ and G cannot act discontinuously at z. Because z was arbitrary and Λ is closed, G does not act discontinuously at any point in $\hat{\mathbb{C}}$. $\square$

We remark that the converse is not true, see Exercise 5.4 below.
We have

Proposition 5.1.3 *If G contains an elliptic element of infinite order then $\Lambda(G) = \hat{\mathbb{C}}$ and G is not discontinuous.*

Proof. If G contains an elliptic of infinite order, we can conjugate G in $\mathcal{M}$ so that the conjugated group contains an element of the form $A(z) = e^{2\pi i\theta}$ for some irrational θ.

Now let p be any point in $\hat{\mathbb{C}} \setminus \{0, \infty\}$ and let U be any neighborhood of p. Since θ is irrational, $A^m(U) \cap U \neq \emptyset$ for infinitely many m and p cannot be in the regular set. Since Λ is closed it also contains zero and infinity. $\square$

Recall that hyperbolic elements are loxodromic.

Proposition 5.1.4 *If p is a fixed point of a loxodromic or parabolic element $A \in G$, then $p \in \Lambda(G)$.*

Proof. Let p be a fixed point of a loxodromic or parabolic A, and let U be a neighborhood of p. Then A^n are distinct for all n and $p \in U \cap A^n(U)$ so that p belongs to $\Lambda(G)$. $\square$

For any $z \in \hat{\mathbb{C}}$, let

$$O(z) = \{A(z)\}_{A \in G}$$

be the orbit of z.

Definition 5.2 *If there are points z whose orbit is finite, G is called an* elementary group.

Examples of elementary groups are the group of affine transformations $z \mapsto z + a$, $a \in \mathbb{C}$, and the group of rotations $z \mapsto e^{2\pi i\theta} z$, $\theta \in \mathbb{R}$. These examples are clearly not discontinuous, although each of them contains discontinuous subgroups. We will discuss discontinuous elementary groups later.

We now characterize the regular and limit sets of non-elementary groups. We need the following lemma.

Lemma 5.1.1 *If* G *is a non-elementary group then* G *must contain infinitely many distinct loxodromic elements.*

Proof. First assume there are no parabolic or loxodromic elements. Because the group is non-elementary, there must be at least two elliptic elements A_1, A_2 in G that do not commute. If they share one fixed point p, however, then the commutator $K = [A_1, A_2] = A_1 A_2 A_1^{-1} A_2^{-1}$ is parabolic with fixed point p, and, if they have no common fixed point, then for n large enough $A_1^n A_2$ is loxodromic (see Exercise 5.1 below), contradicting the assumption.

Thus, if there are no loxodromic elements there must be a parabolic element K whose fixed point p cannot be fixed by every element. Therefore, there is a parabolic L with a different fixed point. Then, by Exercise 5.2 below, for n large enough $[K^n, L]$ is loxodromic.

Let L be a loxodromic element with fixed point p. If A is an element of G such that $p' = A(p) \neq p$, then p' is fixed by the loxodromic element ALA^{-1}. Because the group is non-elementary the orbit $O(p)$ is infinite and there are infinitely many loxodromics. $\square$

For discontinuous groups we can characterize points in $\Omega(G)$ by the following property.

Proposition 5.1.5 *The elements of* G *form a normal family in a neighborhood of* z *if* $z \in \Omega$. *Moreover, if* G *is discontinuous then* Ω *is the largest open set on which they form a normal family.*

Proof. Suppose first that $z \in \Omega$. By Proposition 5.1.1 we can assume that $z = \infty$. Then for some neighborhood U of z, for all but finitely many elements $A \in G$, $A(U)$ is contained in some bounded set and by Montel's theorem they form a normal family.

Next suppose that G is discontinuous and that, for some point $z \in \hat{\mathbb{C}}$ and a neighborhood U of z, the group elements form a normal family on U. It follows that there are sequences A_n of distinct group elements that converge on U either to a constant or to an open map. We claim that any accumulation point of such a sequence has to be a constant.

If not, there is a sequence A_n converging to an open map and the limit function $A(w)$ must be a Möbius transformation. By Weierstrass' theorem, $A_n'(w)$ converges to $A'(w)$ and $A_n''(w)$ converges to $A''(w)$ for all w in U. A simple calculation shows that $\|A_n - A\|$ converges to zero so that by Corollary 1.7.1 G is not discrete. By Proposition 5.1.2, however, this contradicts the hypothesis that G is discontinuous.

Suppose now that $z \in \Lambda$. By the definition of Λ, there are a sequence A_n in G and a sequence of points z_n in a neighborhood U of z such that both z_n and $A_n(z_n)$ converge to z. By the assumption that the group acts normally on U, A_n has to converge to a constant locally uniformly on U and that constant must be z. It follows that, if V is any open subset of U containing z such that $\overline{V} \subset U$, then, for sufficiently large n, $A_n(V) \subset V$. Now consider the sequence A_n^{-1}. By the assumption that the group is normal on U, A_n^{-1} must contain a subsequence converging to some constant. But this is impossible because, for sufficiently large n, $V \subset A_n^{-1}(V)$, and any limit must be an open map. We therefore conclude that $z \in \Omega$. $\square$

An immediate corollary of the proof is

Corollary 5.1.1 *If G is discrete, then the elements of G do not form a normal family at any point of $\Lambda(G)$.*

Now, given a group G, for every z we define $\Lambda(z)$ to be the accumulation set of $O(z)$. That is, $\Lambda(z)$ is the set of all w for which there exists a sequence of distinct $A_n \in G$ such that $A_n(z)$ converges to w. Note that, if z is a fixed point of an element A of infinite order, then A^n are distinct for all n and $z \in \Lambda(z)$.

Proposition 5.1.6 *For any $z \in \hat{\mathbb{C}}$, we have $\Lambda(z) \subset \Lambda(G)$; moreover if G is a discrete non-elementary group then $\Lambda(z) = \Lambda$.*

Proof. Fix $z \in \hat{\mathbb{C}}$. Suppose $z \in U$ and $w \in \Lambda(z)$. By the definition of $\Lambda(z)$ there exists a sequence of distinct maps $A_n \in G$ such that $A_n(z)$ converges to w. Thus, there is an N such that $A_n(z) \in U$ for all $n \geq N$. Therefore, the maps $C_n = A_n \circ A_N^{-1}$ are distinct maps in G and $C_n(U) \cap U \neq \emptyset$. It follows that $\Lambda(z) \subset \Lambda$.

Now suppose G is discrete and non-elementary, $w \in \Lambda$ and $z \in \hat{\mathbb{C}}$. Since G is non-elementary, it is an infinite group, so that $O(z)$ is infinite and $\Lambda(z)$ is non-empty. By the invariance of G, $\Lambda(z)$ is also infinite. Let $\zeta_1, \zeta_2, \zeta_3$ be three distinct points in $\Lambda(z)$. By Corollary 5.1.1 the group elements are not normal on any neighborhood U of w and by Montel's theorem we can find a group element C such that at least one of the points ζ_i, $i = 1, 2, 3$, say ζ_1, belongs to $C(U)$. Then point $C^{-1}(\zeta_1)$ belongs to U and, because $\Lambda(z)$ is closed, we conclude that $w \in \Lambda(z)$ and $\Lambda(z) = \Lambda$. $\square$

An immediate corollary is

Corollary 5.1.2 *If G is infinite then $\Lambda(G)$ is non-empty.*

Proposition 5.1.7 *Let* G *be a non-elementary discrete group.* Λ *is the smallest non-empty* G *invariant closed subset of* $\hat{\mathbb{C}}$.

Proof. By Corollary 5.1.2 Λ is non-empty. Suppose Λ' is closed and invariant, and let $z \in \Lambda'$. Then, by invariance $O(z) \subset \Lambda'$, and, since Λ' is closed, $\Lambda(z) \subset \Lambda'$. But, by Proposition 5.1.6, this implies $\Lambda \subset \Lambda'$. $\square$

An immediate corollary is

Corollary 5.1.3 *If* G *is discontinuous and non-elementary, then* Ω *is the biggest proper open subset of* $\mathbb{C}$ *on which* G *acts invariantly.*

Let us now characterize points of Λ.

Let $\Lambda_0(G)$ be the set of fixed points of loxodromic elements of a non-elementary group G. Then

Proposition 5.1.8 *If* G *is non-elementary then* $\Lambda(G) = \overline{\Lambda_0(G)}$.

Proof. By Lemma 5.1.1 $\Lambda_0(G)$ is not empty. By conjugation, we see that $\Lambda_0(G)$ is invariant under G and by Proposition 5.1.4 we see that $\Lambda_0(G) \subset \Lambda$. The proposition now follows directly from Proposition 5.1.7. $\square$

Now suppose $z \in \Omega(G)$. If for some neighborhood U of z there are n elements $A_0, \dots, A_{n-1}$ of G such that $A_i(U) \cap U \neq \emptyset$ then the same is true for any U' containing U. Therefore let $k(z) \geq 1$ be the smallest integer such that, for every neighborhood U of z, $A_i(U) \cap U \neq \emptyset$ for distinct elements $A_i \in G$, $0 \leq i \leq k(z) - 1$. Set

$$V(G) = \{z \in \Omega(G) \,|\, k(z) > 1\}.$$

Proposition 5.1.9 *The set* $V(G)$ *forms a discrete subset of* $\Omega(G)$*. The points* $p \in V(G)$ *are fixed points of elliptic elements of finite order. All elliptic elements in* G *fixing* $p \in V(G)$ *form a cyclic subgroup of* G *of finite order.*

Proof. If $\Omega(G)$ is empty the statement is vacuous. Suppose p is in the regular set. If p is the fixed point of some $A \neq Id$, then, by Propositions 5.1.3 and 5.1.4, A must be elliptic of finite order n. Thus, for any small neighborhood U_p of p, clearly $A^j(U_p) \cap U_p \neq \emptyset$ for $j = 0, \dots, n-1$ and $p \in V(G)$.

Let $p \in V(G)$, $k = k(p)$, and let U_p be a small enough neighborhood of p such that $U_p \cap A_i(U_p) \neq \emptyset$ for the set A_i, $i = 0, \dots, k(p) - 1$, and for no other element of G. If p is not a fixed point of A_i then $p_i = A_i(p) \neq p$ and there is a neighborhood $U_p' \subset U_p$ such that $A_i(U_p') \cap U_p' = \emptyset$. Therefore p must be a fixed point of all the A_i.

Suppose the second fixed points of A_i and A_j are not the same. Normalizing, it is easy to compute that the group element $K = A_i A_j A_i^{-1} A_j^{-1}$ is parabolic (see Exercise 5.1 below). Since K also fixes the common fixed point of A_i and A_j, it cannot be in the regular set; we assumed, however, that it is. Therefore all the A_i have the same fixed point set and thus commute. The discreteness follows from the definition of $\Omega(G)$. $\square$

The topology on $X = \Omega/G$ is defined by the condition that the projection map $\Omega \to \Omega/G = X$ be continuous. That is, any set in X whose pre-image is open in Ω is open. Since Ω is Hausdorff we have

Proposition 5.1.10 *If G is a discontinuous group the quotient $X = \Omega/G$ is a Hausdorff space.*

Proof. Assume first that $p, q \in \Omega \setminus V(G)$; then, by Proposition 5.1.5, the group elements form a normal family on Ω and so there is a neighborhood U_p of p, disjoint from all its translates, whose orbit accumulates on $\Lambda(G)$. Thus there is at most one translate of q in U_p. So we can shrink U_p so that there are none in U_p. Similarly, we can choose a neighborhood U_q of q such that there are no translates of U_p in U_q and such that U_q is disjoint from all of its translates. With these choices, the orbits of U_p and U_q are disjoint so that their projections to X are disjoint. It follows that $\pi(U_p)$ and $\pi(U_q)$ are disjoint neighborhoods of $\pi(p)$ and $\pi(q)$ respectively.

Now, if u and v are points in $X \setminus \pi(V(G))$, we can apply the above argument to any pair of lifts $\tilde{u}$ and $\tilde{v}$ to get disjoint neighborhoods.

Now suppose $p \in V(G)$. Again by definition, p is the fixed point of an elliptic element A_p of order n and there is a circle in the elliptic net for A_p bounding a disk U_p small enough that $A_p^j(U_p) = U_p$, $0 \le j \le k(p) - 1$, but U_p is disjoint from any other translate of U_p. Similarly, if $q \in V(G)$ we can find U_q invariant under A_q^j, $0 \le j \le k(q) - 1$, but disjoint from all other translates. Thus, shrinking them if necessary, we can assume that the U_p, U_q and all their other translates are disjoint.

It follows that if either or both $p, q \in V(G)$ then there are U_p and U_q such that $\pi(U_p)$ and $\pi(U_q)$ are disjoint neighborhoods of $\pi(p)$ and $\pi(q)$ respectively; note that the projection may be finite to one, but not one to one, on these neighborhoods. We can then argue as above to obtain separating neighborhoods for any pair of points in X. $\square$

In the above proof, we found neighborhoods of points in $X_0 = X \setminus \pi(V(G))$ for which the projection π is a local homeomorphism. The local inverse maps define local coordinate maps on X_0; if z is the local coordinate in Ω, $\pi(z)$ is the local coordinate on X_0; the elements of G are holomorphic so that X_0

has a conformal structure. At the images of points in $V(G)$, the projection is not a local homeomorphism. If n is the order of the elliptic fixing $p \in V(G)$, the coordinate map has the local form z^n and thus cannot be conformal. These coordinate maps define what is called an *orbifold* structure at these points.

Proposition 5.1.2 says that, if $\Omega(G)$ is not empty, then G is Kleinian although Exercise 5.4 below shows the converse is not true. If we restrict to subgroups of Γ or their conjugates, however, discreteness and discontinuity are equivalent. This is because the elements of Γ are isometries of Δ with respect to the hyperbolic metric. We have

Proposition 5.1.11 *If* G *is Fuchsian then it acts discontinuously on its invariant disk.*

Proof.   Assume that G is Fuchsian with invariant disk Δ and suppose there is a point $z \in \Delta$ at which G does not act discontinuously. Then we can find a hyperbolic disk U centered at z of hyperbolic radius r and an infinite sequence of distinct $A_n \in G$ such that $U \cap A_n(U) \neq \emptyset$ for all n. Since the A_n are isometries, the points $w_n = A_n(z)$ must all satisfy $\rho(z, w_n) < 2r$. We can thus find a subsequence that, by abuse of notation, we also denote by w_n, so that $w_n \to w_\infty \in \Delta$. Choose $C \in \Gamma$ so that $C(w_\infty) = z$. Then $CA_n(z) \to z$. Since CA_n form a normal family, some subsequence converges to a limit function D which is open or constant. Since the CA_n are isometries of Δ, D cannot be constant, but is again an isometry with $D(z) = z$. Now, By corollary 1.7.1, G is not discrete. $\quad\square$

Propositions 5.1.2 and 5.1.11 immediately imply

Corollary 5.1.4 *If* G *is a discontinuous subgroup of* Γ *then* $\Omega \supset \Delta$.

This discussion justifies restricting attention for a Fuchsian group only to its invariant disk D and the boundary of the disk ∂D. By Corollary 5.1.4 we have $\Lambda(G) \subset \partial D$. In the literature, if $\Lambda(G) = \partial D$, G is called a *Fuchsian group of the first kind* whereas, if $\Lambda(G) \neq \partial D$, G is called a *Fuchsian group of the second kind*.

Exercise 5.1   Suppose that A_1 and A_2 are elliptic transformations. If they commute, show that $G = <A_1, A_2>$ is finite. Otherwise, show that if they have one fixed point in common then the commutator $[A_1, A_2] = A_1 A_2 A_1^{-1} A_2^{-1}$ is parabolic and if they have no common fixed point then for some n $A_1^n A_2$ is loxodromic. Hint: Normalize so that A_1 has a fixed point at the origin and A_2 has fixed points at ± 1.

Exercise 5.2 Suppose that A_1 and A_2 are parabolic Möbius transformations that do not have the same fixed point. Show that for n large enough $[A_1^n, A_2]$ is loxodromic. Hint: Normalize so that A_1 has a fixed point at the origin and A_2 has a fixed points at ∞.

Exercise 5.3 Show that $\Omega(G)$ is invariant under G.

Exercise 5.4 Let $P = \{z | z = m + in, m, n \in \mathbb{Z}\}$ and let G be the subgroup of $\mathcal{M}$ whose coefficients belong to P. Prove that G is discrete but does not act discontinuously on $\hat{C}$.

Exercise 5.5 Suppose G is a Fuchsian group with invariant disk D. Show that $\Omega \supset \hat{\mathbb{C}} \setminus \bar{D}$ and that therefore $\Lambda \subset \partial D$.

5.2 Discontinuous elementary groups

If G is not discontinuous, its regular set is empty and the limit set is the whole sphere. On the other hand, if G is a discontinuous group the limit set satisfies

Proposition 5.2.1 *The limit set of a group is empty if and only if the group is finite.*

Proof. If G is finite, then by definition $\Omega(G) = \hat{\mathbb{C}}$ and $\Lambda(G) = \emptyset$. If G is infinite, $\Lambda(z)$ is not empty and by Proposition 5.1.6 $\Lambda(z) \subset \Lambda$. $\square$

Suppose now that $\Lambda(G)$ contains a single point. By invariance, this point is fixed by every element of G and G is elementary. We may assume, conjugating the group by a Möbius transformation if necessary, that the point in $\Lambda(G)$ is the point at infinity. Since by assumption $\Omega(G)$ is non-empty, by Proposition 5.1.3 there are no elliptics of infinite order. It follows from Proposition 5.1.4 that fixed points of any non-elliptic elements must belong to $\Lambda(G)$. Since loxodromic and hyperbolic elements have two fixed points there cannot be any such elements in G. Therefore G may contain parabolics fixing infinity. If G has an elliptic element of finite order, one of its fixed points must be infinity and the other fixed point is not in $\Lambda(G)$. We conclude that the elements of G are affine transformations; either they are parabolic with fixed point infinity and so of the form $z + c$, or they are elliptic elements of finite order of the form $a + (z - a)e^{2\pi i\theta}$, θ rational.

If G has no elliptic elements, all its elements are of the form $z + c$ for some collection of $c \in \mathbb{C}$. By Proposition 5.1.2, since G is discontinuous there is an element $z + a$ such that $0 < |a| \le |c|$ for all other elements. Then either

1- Every element is of the form $z + na$ for $n \in \mathbb{Z}$ and we use the notation $G_a = \langle z \to z + a \rangle$ to emphasize the generator. In this case, $\mathbb{C}/G_a$ is a cylinder.

Let $F : \mathbb{C} \to \mathbb{C}$ be given by $z \mapsto z/a$ and write the generator of G_a as $A(z) = z + a$. Set $A' = FAF^{-1} = z + 1$ and $G_1 = \langle z + 1 \rangle$. Then, since F is holomorphic, it induces a holomorphic map on cylinders

$$f : \mathbb{C}/G_a \to \mathbb{C}/G_1.$$

This shows that, up to conformal conjugacy, there is only one cylinder.

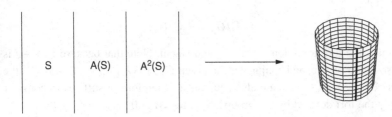

Figure 5.1   The strips S, $A(S)$ and $A^2(S)$ all project to the cylinder

Or

2- The group G_a is a proper subgroup of G. Then, again by Proposition 5.1.2, there is an element $z + b$ not in G_a such that $0 < |a| \le |b| \le |c|$ for all other elements. Write $G_{a,b} = \{z \mapsto z + am + bn \mid m, n \in \mathbb{Z}\}$. If $\Im(b/a) = 0$, one of the elements $z + b \pm a$ would satisfy $|b \pm a| \le |b|$. We can assume, taking inverse elements if necessary, that $\Im(b/a) > 0$. Here, in terms of generators we have $G_{a,b} = \langle z \mapsto z + a, z \mapsto z + b \rangle$. To see that $G_{a,b} = G$, it suffices to note that, for any element $z + c \in G \setminus G_{a,b}$, we can find $n, m \in \mathbb{Z}$ such that $0 \le |c - (an + bm)| < |b|$ (see Exercise 5.7 below). By the choice of b, we

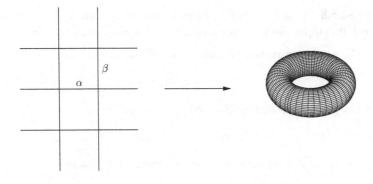

Figure 5.2   Here α and β project to the curves on the torus

must have $c = an + bm$. The group $G_{a,b}$ is called a *lattice group* with lattice $L = \{z \mid z = am + bn,\ m, n \in \mathbb{Z}\}$.

In this case $\mathbb{C}/G_{a,b}$ is a torus.

Now, let $F : \mathbb{C} \to \mathbb{C}$ be given by $z \mapsto z/a$ and write the generators of $G_{a,b}$ as $A(z) = z + a$, $B(z) = z + b$. Set $\tau = b/a$,

$$A' = FAF^{-1} = z + 1 \text{ and } B' = FBF^{-1} = z + b/a = z + \tau$$

and write $G_{1,\tau} = \langle A', B' \rangle$. Then, since F is holomorphic, it induces a holomorphic map

$$f : \mathbb{C}/G_{a,b} \to \mathbb{C}/G_{1,\tau}$$

so that the tori are holomorphically equivalent. Note that because $z \mapsto -z$ is also holomorphic, and conjugates the group $G_{1,\tau}$ to $G_{-1,-\tau} = \langle A'^{-1}, B'^{-1} \rangle$, the tori $\mathbb{C}/G_{1,\tau}$ and $G_{-1,-\tau}$ are also equivalent. Therefore, it suffices to consider only the tori defined by the groups $G_{1,\tau}$, for $\Im \tau > 0$.

Exercise 5.6 If $\Lambda(G)$ contains a single point and G has no elliptic elements, then G contains a subgroup conjugate to $G_a = \langle z \mapsto z + a \rangle$, $a \neq 0$.

Exercise 5.7 Show that, if the group G is a discontinuous group of affine transformations of the form $z + c$ and $z + a$ and $z + b$ are elements of G chosen such that, for all c, $0 < |a| \leq |b| \leq |c|$ and $\Im(b/a) > 0$, then the group $G_{a,b} = \langle z \to z + a, z \to z + b \rangle = G$. That is, show that, for any element $z + c \in G/G_{a,b}$, we can find $n, m \in \mathbb{Z}$ such that $0 \leq |c - (an + bm)| < |b|$.

If there are elliptic elements of finite order in G, their fixed points are infinity and a point of $V(G)$. It is not too difficult to show that the only possibilities are the following. We omit the proofs and leave them as an exercise.

Exercise 5.8 Show that if G is a discontinuous group with $\Lambda(G) = \{\infty\}$ and G contains elliptic elements then it must be conjugate to one of the following:

(i) The group generated by G_1 and $z \mapsto -z$; that is

$$G = \langle z \mapsto z + 1, z \mapsto -z \rangle.$$

(ii) The group generated by $G_{1,it}$ and $z \mapsto -z$; that is

$$G = \langle z \mapsto z + 1, z \mapsto z + it, z \mapsto -z \rangle.$$

(iii) The group generated by $G_{1,i}$, $z \mapsto -z$ and $z \mapsto iz$; that is

$$G = \langle z \mapsto z + 1, z \mapsto z + i, z \mapsto -z, z \mapsto iz \rangle.$$

(iv) The group generated by $G_{1,e^{\frac{2\pi i}{3}}}$ and $z \mapsto e^{\frac{2\pi i}{3}} z$; that is,

$$G = \langle z \mapsto z+1, z \mapsto z+e^{\frac{2\pi i}{3}}, z \mapsto e^{\frac{2\pi i}{3}} z \rangle.$$

(v) The group generated by $G_{1,\frac{1+i\sqrt{3}}{2}}$ and $z \mapsto e^{\frac{2\pi i}{6}} z$; that is,

$$G = \langle z \mapsto z+1, z \mapsto z+e^{\frac{2\pi i}{3}}, z \mapsto e^{\frac{2\pi i}{6}} z \rangle.$$

We turn now to groups where $\Lambda(G)$ contains more than one point.

Theorem 5.2.1 *Let A, B be elements of a subgroup G of $\mathcal{M}$ with exactly one common fixed point. Suppose A is loxodromic. Then G is not discrete.*

Note that by Proposition 5.1.2 this means that G is not discontinuous.

Proof of Theorem. We may assume, as above, that the common fixed point is infinity and the loxodromic transformation (or its inverse) has the form $A(z) = \lambda z$, $|\lambda| > 1$. Since B also has fixed point at infinity it has the form $B(z) = az+b$, and by hypothesis it cannot have zero as a fixed point, so $b \neq 0$.

For any n, the sequence of group elements

$$C_n = A^{-n}BA^n = az + \lambda^{-n}b$$

converges to $C(z) = az$ as n tends to infinity. Therefore, by Corollary 1.7.1, the group is not discrete. $\square$

The following exercises characterize all groups whose limit set contains exactly two points. They are elementary groups.

Exercise 5.9 Deduce from Theorem 5.2.1 that, if $\Lambda(G)$ contains exactly two points, all the elements have these two points as fixed points, with the possible exception of an elliptic element of order 2 that interchanges the points. Note that all these elements commute. Hint: Show that there cannot be any parabolic elements in such a group because they would take the second fixed point to many new points in the limit set.

Exercise 5.10 Show that if $\Lambda(G)$ contains exactly two points, which we assume to be zero and infinity, there is a minimal element $A_0(z) = \lambda_0 z$ such that $|\lambda_0| > 1$ and $|\lambda_0|$ is closest to 1 among all elements of G. That is, either $G = \langle z \mapsto \lambda_0 z \rangle$ or G is generated by $z \mapsto \lambda_0 z$ and one or both of $z \mapsto -z, z \mapsto 1/z$.

Remark. It follows from Exercises 5.6, 5.7, 5.8, 5.9 and 5.10 that, if G is a discontinuous subgroup of $\mathcal{M}$ whose limit set contains zero, one or two points it is an elementary group.

5.3 Non-elementary groups

From now on we will concentrate on non-elementary discontinuous groups.
Proposition 5.1.6 implies

Proposition 5.3.1 *If G is a non-elementary group then $\Lambda(G)$ contains infinitely many points.*

Remark. It follows from Exercise 5.1 and Proposition 5.3.1 that the limit set of a discontinuous group G contains either no points, one point, two points or infinitely many points.

Now by Proposition 5.1.8 it follows that the limit set of a non-elementary group contains infinitely many non-elliptic elements.
We can, however, say more.

Theorem 5.3.1 *If G is a non-elementary group then G contains infinitely many loxodromics with distinct fixed points. Moreover, if G contains a parabolic, it contains infinitely many parabolics with distinct fixed points.*

Proof. Because G is not elementary, by Lemma 5.1.1, it must contain a loxodromic element A. Let p be a fixed point of A. If z is any point in the orbit of p, there exists an element C of G such that $C(p) = z$. The transformation $A_z = CAC^{-1}$ is a loxodromic element in G with a fixed point at z. Since G is not elementary, the orbit of p is infinite. This, together with the fact that each A_z has no more than two fixed points, implies that there are infinitely many A_z's with distinct fixed point sets.

If there is a parabolic element B in G, let q be its fixed point. For every point $z = C(q)$ in the orbit of q let $B_z = CBC^{-1}$. Each B_z is a parabolic element in G with fixed point at z, and thus there are infinitely many parabolics in G with distinct fixed points. $\square$

Exercise 5.11 Let A_1 and A_2 be elements of $\mathcal{M}$ and assume that A_i is not an elliptic of order 2 that interchanges the fixed points of A_j, $i \neq j$. Show that A_1 and A_2 commute if and only if they have the same fixed points.

Exercise 5.12 Show that, except for the special groups, i–v in Section 5.2, the only discrete subgroups of $\{z \mapsto az + b\}$ are the groups $G_1, G_{1,\tau}$ and their conjugates. Hint: Show any non-parabolic elements are elliptic of finite order. Show these fixed points form a lattice in the plane preserved by the group.

Exercise 5.13 Prove that, for any choice of matrix representations of two Möbius transformations A, B, the sign of tr $ABA^{-1}B^{-1}$ is independent of the representation.

Exercise 5.14 Prove that two Möbius transformations A, B have a common fixed point if and only if for any choice of matrix representations $\operatorname{tr} ABA^{-1}B^{-1} = 2$. If $ABA^{-1}B^{-1} = Id$, then A and B have the same fixed point sets. If $ABA^{-1}B^{-1}$ is parabolic, then A and B have different fixed point sets.

Exercise 5.15 Prove that if A and B are parabolics with distinct fixed points then either AB or AB^{-1} (or both) is loxodromic. Hint: Normalize so that the fixed points are zero and infinity.

6
Fuchsian groups

6.1 An historical note

In the eighteenth century mathematicians were busy developing the calculus as we learn it in our introductory courses. They found techniques to integrate rational functions $\int R(x)dx$; in addition to rational functions as solutions, they found they needed to introduce the logarithm function,

$$\log x = \int_1^x \frac{dt}{t}.$$

The next step was to attack integrals of algebraic functions of the form

$$\int R(\sqrt{ax^2 + bx + c})dx. \tag{6.1}$$

They found that they could solve these by introducing the inverse trigonometric functions,

$$\arcsin x = \int_0^x \frac{dt}{\sqrt{1-t^2}} \quad \text{and} \quad \arctan x = \int_0^x \frac{dt}{1+t^2}.$$

Replacing the quadratic in formula (6.1) by a cubic made the problem much harder. These new integrals arose, for example, in computing the arclength of an ellipse and are called elliptic integrals. N. Abel (1802–1829) and C. Jacobi (1804–1851) developed two approaches to dealing with elliptic integrals. The first was to consider the inverses of elliptic integrals. This is analogous to considering the inverse trigonometric functions. The second was to replace the real variable by a complex variable.

The trigonometric functions are periodic, and have a natural extension as functions on the complex plane. The inverse elliptic integrals, called elliptic functions, have two basic periods provided we consider them as functions of a complex variable,

$$E(z+a) = E(z) \text{ and } E(z+b) = E(z), \quad \Im\left(\frac{b}{a}\right) \neq 0.$$

96

It follows, then, that for all integers, m, n,

$$E(z + ma + nb) = E(z).$$

Thus the periods form a group consisting of lattice points in the plane of the form $na + mb$, and the elliptic functions are invariant under the group of Euclidean isometries consisting of elements $z \mapsto z + na + mb$. We will present a short discussion of this approach in the Appendix.

Integrals whose integrands are more complicated algebraic functions led B. Riemann (1826–1866) to develop the theory of Riemann surfaces, but he did not look for inverse functions with periodicities for these integrals. It was H. Poincaré (1854–1912) who found the proper generalization of periodicity. The mathematician H. Fuchs had introduced him to certain second order differential equations whose coefficients had poles. Finding solutions to these equations is equivalent to evaluating the integrals of algebraic functions. The local solutions to these equations change as you make a loop around the pole. Fuchs realized that one should consider quotients of these local solutions because if you write a solution as a linear combination of two normalized solutions, as you go around the pole, the quotient of two solutions changes by a Möbius transformation. Paths around different poles yield different Möbius transformations and the collection of them forms a group called the *monodromy group* of the equation. It was Poincaré who realized that the quotient function was really the inverse of a function defined on a surface associated to the group; that is, it took the same value at points congruent under the elements of the group. In [54], Poincaré writes that, as he stepped onto a bus in the midst of a trip, he suddenly realized that these transformations were just the isometries of hyperbolic geometry and that, therefore, the inverse function defined a hyperbolic structure on the surface. He named these monodromy groups Fuchsian groups in honor of Fuchs (see also [59]).

6.2 Fundamental domains

Let us consider the elementary group G_1 all of whose elements are of the form $A_n(z) = z + n$, $n \in \mathbb{Z}$. We saw in Chapter 5 that its limit set consists of the single point at infinity and that it acts discontinuously on all of $\mathbb{C}$.

Now suppose we consider the strip $S = \{z = x + iy \mid 0 \le x < 1\}$. It is easy to see that $A_n(S) \cap A_m(S) = \emptyset$ unless $m = n$. That is, for any $z \in S$, all its images under G are outside S. If we make S any bigger by including any more points in $\mathbb{C}$ this will no longer be true. The strip contains one and only one representative of each equivalence class of the quotient $X = \mathbb{C}/G$.

This construction tells us explicitly how to define the Riemann surface structure on the quotient. (Recall Proposition 5.1.10.) If we glue the vertical sides together we obtain a sphere X with two points removed. We get a conformal structure at the images of interior points of S directly by projection; a neighborhood in S projects to a neighborhood in X. To make a neighborhood of a point iy on the boundary of S, take small disks $U \subset \mathbb{C}$ of iy and $V = 1 + U$ of $1 + iy$. The neighborhood of iy on X is then $(U \cap S) \cup (V \cap S)$. X with this structure is thus a Riemann surface. In this example we actually already knew how to find the structure on X because we could write down the universal covering projection $e^{2\pi i z}$. Thus the structure we obtained for X from the strip with its edges identified is holomorphically equivalent to the usual structure of $\mathbb{C} \setminus \{0\}$. This motivates us to define

Definition 6.1 *Let G be a discontinuous subgroup of $\mathcal{M}$. A domain $F \subset \Omega$ is called a* fundamental domain *for G if the following two conditions hold:*

$$A(F) \cap F = \emptyset \text{ for all } A \in G, A \neq Id;$$

$$\bigcup_{A \in G} \overline{A(F)} = \Omega.$$

A set that satisfies only the first condition is called a partial fundamental domain.

The interior of the strip S in the example above is a fundamental domain for the group G_1 of translations.

Let's consider another simple example. Let $G_{1,i}$ be the elementary group all of whose elements are of the form $A_{m,n}(z) = z + m + ni$, $m, n \in \mathbb{Z}$, and let

$$F = \{z = x + iy \,|\, 0 < x < 1, 0 < y < 1\}.$$

Then it is easy to check that the square F is a fundamental domain.

Glue the vertical and horizontal boundaries of F to each other and adjoin the glued boundaries to F to form a surface X. It is easy to see that gluing the horizontal boundaries we get a cylinder whose boundaries are the vertical edges which have become circles. Adding these circles and gluing them together we get a torus. To find a neighborhood for each point inside the square, we just use its neighborhood in $\mathbb{C}$. For points on the boundary sides we take semi-disks as we did in the example above. For corner points, we take quarter disks and glue them together.

Note that there are many fundamental domains for a given group. In this example, the parallelogram

$$F' = \{z = x + (1+i)y \,|\, 0 < x < 1, 0 < y < 1\}$$

is also a fundamental domain. The basic periods for this domain are $(1, 1 + i)$ as opposed to $(1, i)$ for F. In fact, we can take as basic periods any pair $(m_1 + n_1 i, m_2 + n_2 i)$ such that $m_1 n_2 - m_2 n_1 = \pm 1$ (see Exercise 6.1 below). The conformal structure on the Riemann surface obtained by gluing the sides is the same for any of these fundamental domains.

In each of these examples we also obtain a natural metric on the surface by requiring that the projection map $\pi : \mathbb{C} \to \mathbb{C}/G$ be a local isometry with respect to the Euclidean metric.

Before we discuss non-elementary groups we give an example of a fundamental domain for an elementary group whose limit set contains two points. Let G be the elementary group all of whose elements are of the form $A_n(z) = \lambda^n z$, $\lambda > 1$, $n \in \mathbb{Z}$. We can take as fundamental domain the annulus

$$F = \{z \mid 1 < |z| < \lambda\}.$$

If we add the boundary curves and glue the boundary $|z| = 1$ to the boundary $|z| = \lambda$, we obtain a torus X.

Although $\Omega(G) = \hat{\mathbb{C}} \setminus \{0, \infty\}$, G leaves the upper half plane $\mathbb{H}$ invariant and therefore can be thought of as a Fuchsian group. Also, by Corollary 5.1.4, the invariant disk D for any Fuchsian group is a subset of the set of discontinuity. Since when we work with a Fuchsian group we concentrate on its action on its invariant disk we define

Definition 6.2 *Let G be a Fuchsian group with invariant disk D. A domain $F \subset D$ is called a* fundamental domain *for G if the following two conditions hold:*

$$A(F) \cap F = \emptyset \text{ for all } A \in G, A \neq Id;$$

$$\bigcup_{A \in G} \overline{A(F)} = D.$$

A set that satisfies only the first condition is called a partial fundamental domain.

Therefore, if we consider the group G above, whose elements are $A_n(z)$, as a Fuchsian group with invariant disk $\mathbb{H}$, we can take as a fundamental domain the semi-annulus

$$F_R = \{z \mid z \in \mathbb{H}, 1 < |z| < \lambda\}.$$

If we add the boundaries $\{z \mid |z| = 1\} \cap \mathbb{H}$ and $\{z \mid |z| = \lambda\} \cap \mathbb{H}$, and glue them together, we obtain an open annulus X.

In the first two examples we defined a metric on the quotient by projecting the Euclidean metric. We could do this because the group elements were isometries in this metric. In the third and fourth examples the group elements are not isometries in the Euclidean metric so we cannot project it to the quotient. In the fourth example, however, the group elements $A_n(z) = \lambda^n z$ are isometries in the hyperbolic metric and we can project this metric locally to obtain a metric on X. We discuss this in detail in Chapter 7.

We can use this metric to characterize the boundary of the quotient X of the fourth example. Since the group G in that example acts discontinuously on $\hat{\mathbb{R}} \setminus \{0, \infty\}$, the two boundary circles of the quotient $\overline{X} = \overline{\mathbb{H}}/G$ come from the boundary intervals of the semi-annulus F_R on the real axis. They form "the boundary at infinity" of X since their points are at infinite hyperbolic distance from any point in F_R.

We call a fundamental domain F together with all its translates a *tessellation* or *tiling* of the set of discontinuity. In the first and second examples, we have a tiling of the Euclidean plane by strips and parallelograms. In the third example the plane with the origin removed is tiled by the annuli and in the last example $\mathbb{H}$ is tiled by the semi-annuli.

Exercise 6.1 Show that the domain

$$F_{m_1,m_2,n_1,n_2} = \{z \mid z = s(m_1 + n_1 i) + t(m_2 + n_2 i), 0 < s < 1, 0 < t < 1\}$$

where $m_1 n_2 - m_2 n_1 = \pm 1$ is a fundamental domain for the elementary group $G_{1,i}$.

Exercise 6.2 Show that the map $\pi(z) = e^{iz}$ is a universal covering map from the plane onto the punctured plane $\mathbb{C} \setminus \{0\}$. Make the punctured plane $\mathbb{C} \setminus \{0\}$ into a metric space by assuming that $\pi(z)$ is a local isometry with respect to the Euclidean metric in the plane $\mathbb{C}$. That is, $\mathbb{C} \setminus \{0\}$ inherits the Euclidean metric from its covering space. Evaluate the distance between $\frac{1}{2}$ and $\frac{1}{3}i$ in $\mathbb{C} \setminus \{0\}$ with respect to this metric.

Exercise 6.3 Show that the map $\pi(z) = e^{iz}$ is a universal covering map from the upper half plane onto the punctured disk $\Delta^* = \Delta \setminus \{0\}$. Make the punctured plane $\mathbb{C} \setminus \{0\}$ into a metric space by assuming that $\pi(z)$ is a local isometry with respect to the Euclidean metric in the upper half plane. That is, Δ^* inherits the Euclidean metric from its universal covering space. Evaluate the distance between $\frac{1}{2}$ and $\frac{1}{3}i$ in Δ^* with respect to this metric. Now define another metric on Δ^* by assuming that $\pi(z)$ is a local isometry with respect to the hyperbolic metric in the upper half plane. Evaluate the distance between $\frac{1}{2}$ and $\frac{1}{3}i$ with respect to this metric. Which of the two metrics is complete?

Exercise 6.4 Show that the map $\pi(z) = \exp\left(-\frac{i \log \lambda \log z}{\pi}\right)$ is a universal covering map from the upper half plane onto the annulus $A = \{w|1 < |w| < \lambda\}$. Make the annulus A into a metric space by assuming that $\pi(z)$ is a local isometry with respect to the hyperbolic metric in the upper half plane. That is, A inherits the hyperbolic metric from its covering space. Evaluate the distance between $\frac{1+\lambda}{2}$ and $\left(\frac{1+2\lambda}{3}\right)i$ in A with respect to this metric.

6.3 Dirichlet domains and fundamental polygons

There are many different ways to find a fundamental domain for a Fuchsian group, each with advantages and disadvantages. There are several very useful conditions that a fundamental domain can satisfy that make it easier to work with. The first useful condition is hyperbolic convexity defined in Definition 2.4. We repeat the definition here.

Definition 6.3 *A subset D of Δ (or $\mathbb{H}$) is called* hyperbolically convex *if, for every pair of points $z, w \in D$ (or $\mathbb{H}$), the hyperbolic geodesic joining z to w lies inside D. If the context is clear, we drop the adjective and just say the set is* convex.

Simple examples of convex sets are a half plane and a disk. They are convex in both the Euclidean and hyperbolic senses. It is easy to check that the intersection of convex sets is convex (see Exercise 6.6 below).

One construction of a convex fundamental domain that is simple to explain is the following. We will describe it in the disk model Δ for definiteness but the construction works just as well for the upper half plane model and we will give examples in $\mathbb{H}$ as well as in Δ.

Pick a point $z_0 \in \Delta$ that is not a fixed point of any element of a Fuchsian group G, and, for each $A \in G \setminus \{Id\}$, form the open set of points

$$D_A = \{z|\rho(z_0, z) < \rho(A(z_0), z)\}.$$

The equation $\rho(z_0, z) = \rho(A(z_0), z)$ defines a geodesic γ_A in Δ that is the perpendicular bisector of the geodesic segment joining z_0 and $A(z_0)$. Note that γ_A is the boundary of the set D_A so that D_A is a half plane in Δ.
Set

$$D = \bigcap_{A \in G} D_A.$$

We call $D = D(z_0)$ the *Dirichlet domain* for G at z_0. It is non-empty since it contains z_0 and, as an intersection of half planes, it is convex.
We have

Proposition 6.3.1 *D is a fundamental domain for G.*

Proof. To see that D is a domain, note first that, for any $z \in D$, the geodesic segment joining z_0 to z belongs to D so that D is connected. Now suppose it is not an open set. Then there are a point $z^* \in D$ and a sequence $z_n \notin D$ with $z_n \to z^*$. By moving each z_n slightly we may assume that there is some A_n such that

$$\beta'_n = \rho(z_n, A_n(z_0)) < \rho(z_n, z_0) = \alpha'_n$$

and

$$\beta_n = \rho(z^*, A_n(z_0)) > \rho(z^*, z_0) = \alpha_n.$$

Because $z_n \to z^*$ the α'_n are bounded. Since G is Fuchsian and acts discontinuously on Δ, we can choose a subsequence of the A_n so that they are all the same for large enough n. Letting n tend to infinity, $\beta_n = \beta > \alpha = \alpha_n$, but $\beta'_n \to \beta$ and $\alpha'_n \to \alpha$ imply $\beta \le \alpha$.

To see that D is a fundamental domain, we have to check both conditions in the definition. For the first condition, let z_1 belong to D. Then, for $A \in G \setminus \{Id\}$, since A is an isometry,

$$\rho(z_0, A(z_1)) = \rho(A^{-1}(z_0), z_1) > \rho(z_0, z_1) = \rho(A(z_1), A(z_0)).$$

Thus $A(z_1) \notin D$ and the first condition holds.

Next suppose $z \in \Delta \setminus 0(z_0)$. Because G is Fuchsian and is discontinuous at every point of Δ, we can find some element A such that

$$0 < \rho(z_0, A(z)) \le \rho(z_0, B(z))$$

for all $B \in G$. We claim that $A(z) \in \overline{D}$. It is sufficient to prove that every interior point ζ on the geodesic segment g joining z_0 to $A(z)$ lies in $\overline{D}$. If some such $\zeta \in g$ is not in $\overline{D}$, by moving ζ slightly toward $A(z)$ if necessary, we may assume there is an element $C \in G \setminus \{Id\}$ such that $\rho(z_0, \zeta) > \rho(C(z_0), \zeta)$. Then

$$\rho(C(z_0), A(z)) \le \rho(C(z_0), \zeta) + \rho(\zeta, A(z))$$
$$< \rho(z_0, \zeta) + \rho(\zeta, A(z)) = \rho(z_0, A(z)).$$

Therefore

$$\rho(z_0, C^{-1}A(z)) = \rho(C(z_0), A(z)) < \rho(z_0, A(z)),$$

contradicting the definition of A. $\square$

Remark. Suppose that h is a Möbius transformation mapping Δ onto $\mathbb{H}$ and that G is a subgroup of Γ. Then h defines an isomorphism $\theta_h : G \to G_h \subset PSL(2, \mathbb{R})$ given by $\theta_h : A \mapsto hAh^{-1}$ for all $A \in G$. Since h maps hyperbolic half planes in Δ to hyperbolic half planes in $\mathbb{H}$, if $D(z_0)$ is a Dirichlet domain for G, then $D(h(z_0)) = h(D(z_0))$ is a Dirichlet domain for G_h.

Before we continue with our general discussion we look at three examples. We leave the proofs to the exercises.

Example 1 The group $G = PSL(2, \mathbb{Z}) \subset PSL(2, \mathbb{R})$ acting on $\mathbb{H}$ has entries in its matrices that are not only real but integers:

$$A = \frac{az+b}{cz+d}, \ a, b, c, d \in \mathbb{Z}, \ ad - bc = 1.$$

Let $z_0 = 2i$. The domain

$$D(2i) = \left\{ z = x + iy \mid -\frac{1}{2} < x < \frac{1}{2}, |z| > 1 \right\}$$

is the Dirichlet domain at $2i$ for G. (See Exercise 6.8 below and Figure 6.1.)

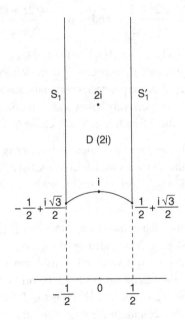

Figure 6.1   The domain D(2i)

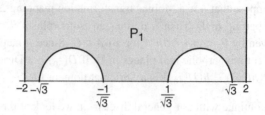

Figure 6.2   Fundamental domain for Example 2

Example 2 Consider the group G whose generators are

$$A = \frac{2z + \sqrt{3}}{\sqrt{3}z + 2} \text{ and } B = z + 4.$$

Let s_1 be the geodesic in $\mathbb{H}$ with endpoints $(-\sqrt{3}, -\frac{1}{\sqrt{3}})$ and let s'_1 be the geodesic with endpoints $(\frac{1}{\sqrt{3}}, \sqrt{3})$. Let s_2 be the geodesic (vertical line) joining $(-2, \infty)$ and s'_2 the geodesic joining $(2, \infty)$. Then the polygon P_1 of Figure 6.2 bounded by these four geodesics is the Dirichlet domain $D(i)$. (See Exercise 6.9 below.)

Example 3 Let $G \subset \Gamma$ be the group acting on Δ whose generators are

$$A = \frac{\sqrt{2}z + 1}{z + \sqrt{2}} \quad \text{and} \quad B = \frac{\sqrt{2}z + i}{-iz + \sqrt{2}}.$$

Let $\mu = \frac{1}{2} \operatorname{arc cosh} \sqrt{2}$. Let $s_1, s'_1 = A(s_1)$ be the geodesics perpendicular to the real axis at hyperbolic distance μ, respectively to the left and right of the origin. Similarly, let $s_2, s'_2 = B(s_2)$ be the geodesics perpendicular to the imaginary axes at hyperbolic distance μ, respectively below and above the origin. Then $D(0)$ is the domain bounded by these four lines. (See Exercise 6.10 below.)

A second useful condition for fundamental domains is local finiteness. All the fundamental domains in the examples above satisfy it, as do the Dirichlet domains (see Proposition 6.3.2 below). Before we give the definition, we look at another example.

Example 4 Let G be the elementary group acting on $\mathbb{H}$ generated by $A(z) = \frac{3}{2}z$. Let F be the domain in $\mathbb{H}$ bounded by the geodesic s with endpoints 0 and 1 and the geodesic $s' = A(s)$ with endpoints 0 and $\frac{3}{2}$.

For any $z \in F$, draw the Euclidean ray from the origin through z. This line intersects ∂F in two points and clearly $(\frac{3}{2})^n z$ is not in F for all $n = \pm 1, \pm 2, \dots$ so that F satisfies the first condition for a fundamental domain. Note that the translates $A^n(s')$ accumulate onto the imaginary axis as n goes to infinity.

In fact, A leaves both the right and left quadrants invariant so that no point in the left half can be congruent to a point in F. It follows that F does not satisfy the second condition for a fundamental domain and is only a partial fundamental domain.

Note also that F can be obtained as a limit of fundamental domains as follows. Let P_0 be the domain between the geodesic joining $-\frac{3}{2}$ and $\frac{3}{2}$ and the geodesic joining -1 and 1. Then set P_i, $i = 1, 2, \ldots$, to be the domain between the geodesic joining $A^{-i}(-\frac{3}{2})$ and $\frac{3}{2}$ and the geodesic joining $A^{-i}(-1)$ and 1. By Exercise 6.11 below each P_i is a fundamental domain and, since the origin is an attracting fixed point for A^{-1}, the domains P_i converge to F.

This pathology leads us to the definition of local finiteness.

Definition 6.4 *A partial fundamental domain or a fundamental domain F is locally finite if, for any compact set $K \subset \Delta$, only finitely many translates of F meet it. Said more precisely, $A(F) \cap K = \emptyset$ for all but finitely many $A \in G$.*

In the next example we consider a non-elementary group that contains the group of Example 4 as subgroup. We find a non-locally-finite (full) fundamental domain for this group as follows.

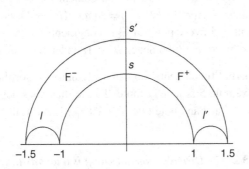

Figure 6.3 The fundamental domain F_0

Example 5 Let F_0 be the domain in $\mathbb{H}$ bounded by the geodesic s joining -1 and 1, the geodesic s' joining $-\frac{3}{2}$ and $\frac{3}{2}$, the geodesic l joining $-\frac{3}{2}$ and -1 and the geodesic l' joining 1 and $\frac{3}{2}$. Let $A(z) = \frac{3}{2}z$ and let $B(z) = (5z+6)/(4z+5)$ so that $A(s) = s'$ and $B(l) = l'$. It is not hard to check, and will follow from Poincaré's theorem (see Section 6.5), that F_0 is a fundamental domain for $G = \langle A, B \rangle$.

Let F^- be the intersection of F_0 with the left half plane and let F^+ be its intersection with the right half plane. Then, $F_1 = F^+ \cup B(F^-) \cup l'$ is a

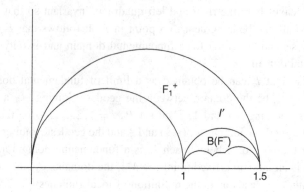

Figure 6.4   The non-locally-finite full fundamental domain $F_1^+ \cup B(F^-) \cup l'$ of
Example 5

fundamental domain for G because we have replaced F^- with a congruent partial fundamental domain. We add in l' so that the domain is connected.

Next let F_1^+ be the domain bounded by the geodesic joining 0 and $\frac{3}{2}$, the geodesic joining 0 and 1 and the geodesic l'. Now (see Figure 6.4) $F = F_1^+ \cup B(F^-) \cup l'$ is also a fundamental domain for G since, for any ray η in the first quadrant $A(\eta \cap (F_1^+ \backslash F^+) = A(\eta \cap (F^+ \backslash F_1^+)$, no two points in F_1^+ are congruent and every point in $\mathbb{H}$ is congruent to exactly one point in $\overline{F^+}$ or to some point in $\overline{F^-}$ (and hence $\overline{B(F^-)}$) under some element of G.

Neither the partial fundamental domain of Example 4 nor the fundamental domain F of Example 5 is locally finite. To see this, consider the disk K of radius $1/2$ centered at i and note that $K \cap A^n(F_1^+) \neq \emptyset$ for all $n > 0$.

We have

Proposition 6.3.2 *The Dirichlet domain $D(z_0)$ is a locally finite fundamental domain.*

Proof.   To see that the Dirichlet domain $D(z_0)$ satisfies the local finiteness condition, it is sufficient to take K as a closed hyperbolic disk of radius r centered at z_0. Since the group G is discontinuous, by Proposition 5.1.6, there are only finitely many A for which $\rho(z_0, A(z_0)) < 2r$. For all others, $A(D(z_0)) \cap K = \emptyset$.   □

In the examples of fundamental domains of elementary groups in Section 6.2 above, the boundary curves of each fundamental domain are Euclidean geodesics that we call sides of F. There are an even number of sides and they are paired by group elements. For example, in the square fundamental domain of the group $G_{1,i}$, the vertical sides are paired: $z + 1$ takes the left side s_v to the right side s_v' and $z - 1$ takes s_v' to s_v. Similarly the horizontal sides are paired: $z + i$ takes the bottom side s_h to the top side s_h' and $z - i$ takes s_h' to s_h. If we think of the fundamental domain as a tile, the side pairings tell us how to put the tiles down on the plane. In addition they tell us how to glue the fundamental domain up to get a Riemann surface.

Recall that we call a Euclidean domain whose boundary consists of straight lines a polygon and we call the lines in the boundary sides. Similarly, in subsection 2.1.2, a hyperbolic domain whose sides are geodesics was called a hyperbolic polygon or more simply a polygon if we are in the hyperbolic plane.

As we saw in subsection 2.1.2, a hyperbolic polygon may have sides of infinite length; such a side meets $\partial \Delta$ (or $\partial \mathbb{H}$) in a right angle. If two infinite sides meet at a point on $\partial \Delta$ (or $\partial \mathbb{H}$), they must be tangent there and so meet at a zero angle; such a point is called *an ideal vertex* of the polygon. A polygon may have an infinite side that does not meet another infinite side at its boundary point; such a side is called a *free side*. In this case, there may be arcs of $\partial \Delta$ (or $\partial \mathbb{H}$) in the set theoretic boundary of the polygon. We call the endpoint of a free side a *free vertex*. It is also possible for a polygon to have infinitely many sides.

It is useful to work with fundamental domains that are polygons. In fact, all the examples we have considered so far are polygons and all Dirichlet domains are polygons (see Exercise 6.7 below).

Note that, if z_0 is the fixed point of an elliptic element, it cannot be in the interior of a fundamental polygon (see Exercise 6.12 below). We call vertices of a polygon inside Δ that are not fixed points of elliptic elements *accidental vertices*.

A useful condition for hyperbolic polygons is

Definition 6.5 *A polygon F is called a* side-paired polygon *for a group G if it is the interior of a polygon whose sides are paired by group elements. Moreover, if s and s' are paired by $A \in G$, then $s' = A(s)$ and $\overline{A(F)} \cap \overline{F} = s'$. If A is not an elliptic of order 2, $s \neq s'$. If it is elliptic of order 2, its fixed point lies on a side. In this case, we consider the fixed point as a vertex and the two segments on either side of the fixed point as the sides s, s'; then $\overline{A(F)} \cap \overline{F} = s \cup s'$.*

Note that a side-paired polygon must have an even number of sides. We can think of a side-paired fundamental polygon as a hyperbolic tile. Keeping track of the side pairings is useful because it tells us how to tile the hyperbolic

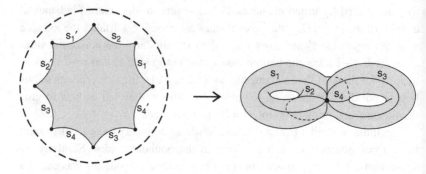

Figure 6.5   A polygon for a genus-2 surface S. The projections of the boundary
of the polygon are shown on S

plane with copies of this tile. It also tells us how to glue the polygon together
to make the quotient surface and makes it easier to determine the topological
type of the surface.

Note that the polygon of Example 5 is NOT a side-paired polygon; it has
five sides.

Notation: In the rest of this chapter we will reserve the notation A for side
pairing elements of the group. Other elements will be denoted by other letters.
In particular, elements that are products of side pairing elements, that is, words
in the side pairing elements, will be denoted by W.

The side pairings of F determine a labeling of the sides of F: the sides
paired by A_i are labeled s_i and s_i'. This side labeling of F determines a side
labeling for all the translates of F under G. For example, $A_1(F)$ and F have
a common side; its label in F is s_1' but its label in $A_1(F)$ is s_1. The vertices at
the ends of the common side have different labels in F and $A_1(F)$; the vertex
labeled v_1 in F is labeled v_0 in $A_1(F)$.

An important property of the side pairings is

Proposition 6.3.3 *The side pairings of a side-paired locally finite convex
fundamental polygon F generate G.*

Proof.   Let z_0 be a point in F and let z be any point in Δ. Draw a geodesic γ
from z_0 to z. By Exercise 6.15 below, perturbing z_0 slightly we can assume that
γ avoids all vertices of F and its translates. The path γ thus crosses sides of
various translates $F_1, \ldots, F_m$ of F as it goes from z_0 to z. Because F is locally
finite, γ crosses only finitely many translates, and, because F is convex, γ
doesn't re-enter F once it leaves. Using the label $s_{j(k)}$ of the side on which γ
enters F_k, we can check that z lies in the translate $A_{j(1)}A_{j(2)} \cdots A_{j(m)}(F)$.

Given any element $W \in G$, choose $z_0 \in F$ and $z = W(z_0)$ so that the geodesic segment γ joining them doesn't pass through a vertex. Since z_0 is in the interior of F, by Exercise 6.12 below, it is not the fixed point of an elliptic element of G and, reading off the side pairings from the side labels along γ, we see W can be written as a product of the side pairings. $\square$

Note that we do not claim the recipe for finding the expression for the word W in terms of generators is unique. In fact, it depends on the choice of z_0 and γ. Moving γ from one side of a vertex to another changes the expression. This non-uniqueness comes from relations in the generators. We will come back to this when we discuss cycles.

By Exercise 6.7 below, the Dirichlet domain D is a polygon. We want to show its sides are identified in pairs. Suppose z is in the interior of a side of D. Then, for some A, $z \in \gamma_A$ and $\rho(z_0, z) = \rho(A(z_0), z)$. Moreover, for any $B \neq A$, such that $B \neq Id$, we have $\rho(z_0, z) < \rho(B(z_0), z)$. Then

$$\rho(z_0, A^{-1}(z)) = \rho(A(z_0), z) = \rho(z_0, z) = \rho(A^{-1}(z_0), A^{-1}(z)).$$

Now for $B \neq A^{-1}$, $B \neq Id$,

$$\rho(B(z_0), A^{-1}(z)) = \rho(AB(z_0), z) > \rho(z_0, z) = \rho(z_0, A^{-1}(z)).$$

It follows that $A^{-1}(z)$ also lies on a side and that the side is $\gamma_{A^{-1}} = A^{-1}(\gamma_A)$. Note also that $A(\gamma_{A^{-1}}) = \gamma_A$.

In the remainder of our discussions we will restrict the use of the name fundamental polygon.

Definition 6.6 *A* fundamental polygon *is a fundamental domain that satisfies all three conditions; that is, it is a convex, locally finite, side-paired polygon.*

Exercise 6.5 Prove that the domain

$$F = \{z = x + (1 + 3i)y \mid 0 < x < 1, 0 < y < 1\}$$

is a fundamental domain for the group $G_{1,i}$ with elements $A_{m,n}(z) = z + m + ni$, $m, n \in \mathbb{Z}$. Show it is locally finite.

Exercise 6.6 Prove that the intersection of convex sets is convex in both the Euclidean and hyperbolic contexts.

Exercise 6.7 Prove that a Dirichlet domain is a polygon. Moreover, a point $z \in \Delta$ is an interior point of a boundary side of $D(z_0)$ if and only if $\rho(z_0, z) \leq \rho(A(z_0), z)$ for all $A \in G$ with equality for exactly two A in G. Similarly, $z \in \Delta$ is a vertex of $D(z_0)$ if and only if $\rho(z_0, z) \leq \rho(A(z_0), z)$ for all $A \in G$ with equality for exactly three A in G.

Exercise 6.8 Prove that the domain in Example 1 is a Dirichlet domain at $2i$ for the group G generated by $A = z + 1$ and $B = -1/z$; A pairs the vertical sides and B maps the side $|z| = 1$ to itself with fixed point at i.

Exercise 6.9 Prove that the domain in Example 2 is a Dirichlet domain at the point i for the group G. Note that A is hyperbolic with axis the unit circle. Conclude that G is Fuchsian.

Exercise 6.10 Prove that the domain in Example 3 is a Dirichlet domain at the origin for the group G. Note that the axes of A and B are the real and imaginary axes respectively. Conclude that G is Fuchsian.

Exercise 6.11 Prove that each of the domains P_i in Example 4 is a fundamental domain for the group by comparing P_i with P_{i+1}.

Exercise 6.12 Prove that, if z_0 is the fixed point of an elliptic element of the Fuchsian group G, it cannot be in the interior of a fundamental polygon.

Exercise 6.13 Show that if s is a side of a fundamental polygon, and v is an endpoint of s inside the unit disk, then there exists exactly one other side coming out of v.

Exercise 6.14 Let F_0 be the fundamental polygon of Example 5. Choose a point $a = it$, $1 < t < \frac{3}{2}$, in F_0 and draw the geodesics γ_1 and γ_2 from -1 and 1 to a respectively. Let T be the triangle in F_0 bounded by γ_1, γ_2 and the side s joining ± 1. Show that $(F_0 \setminus \overline{T}) \cup A(T) \cup s'$ is a non-convex fundamental polygon for G.

Exercise 6.15 Use the convexity to show that a fundamental polygon can have only countably many sides.

6.4 Vertex cycles of fundamental polygons

Suppose now that we have a side-paired locally finite convex fundamental polygon F for the group G. By ∂F we mean its boundary in the hyperbolic plane Δ (or $\mathbb{H}$).

We can use the side pairing transformations to define an equivalence relation on $\overline{F}$ as follows. We set $z \sim w$ if there is a side pairing transformation A such that $A(z) = w$. The point here is that $\overline{F}/\sim$ gives us one representative for each point in Ω/G; that is, each point in F and each point on ∂F contribute one point to the quotient.

Points on ∂F that are not interior to sides of F are points where two sides meet. These are the vertices of the polygon. We consider a point on a side that is fixed by an element of order 2 as a vertex. The equivalence relation on $\overline{F}$ defines an equivalence relation on the vertices:

Definition 6.7 *An equivalence class of (finite) vertices of $\overline{F}/\sim$ is called a cycle of vertices of the polygon.*

We can associate a *cycle transformation* to each of the cycles. This is a word in the generators and is defined up to cyclic permutation. To find the transformation we choose one of the vertices in the cycle, say v_0. Draw a circle γ enclosing v_0, but not any other vertex, that passes through each of the sides having v_0 as an endpoint. By Exercise 6.13, there are two such sides of F. Since the polygon is locally finite, there are finitely many translates of F that meet at v_0 and each contributes a side intersecting γ. As γ passes through each of the translates meeting at v_0, we write out a word using the side labeling. Labeling the translates F_k, as we traverse γ counterclockwise, we obtain the words $W_0 = Id$, $W_k = A_1 A_2 \ldots A_k$, $k = 1, \ldots, n$, with $F_k = W_k(F)$. Since γ is closed we see that $F_n = F$ and $W_n = Id$. The word W_n is called a *cycle relation* or a *relator for the group*.

Applying the transformations W_k^{-1} to v_0, we obtain the vertices $v_k = W_k^{-1}(v_0)$ in F equivalent to v_0. It follows from Exercise 6.16 below that v_0 is elliptic if and only if all the v_k in the cycle are elliptic. The cycles are called *elliptic* or *accidental* as v_0 is elliptic or accidental.

If the cycle is elliptic, the vertices v_k, $k = 1, \ldots, n-1$, are not all distinct and, for some $k < n$, $W_k(v_0) = v_0$. Let $C = W_j$ be the first time this happens. Then C is called the cycle word and $C^2(v_0) = W_{2j}(v_0) = v_0$. Since G is discontinuous at v_0, C is elliptic of finite order q where $qj = n$ and $C^q = Id$.

Let $V = \bigcup_0^{j-1} W_k(F)$ denote the union of the first j translates. The arc of γ in V that has passed through only the first j translates around v_0 is not a closed curve and V is not a full neighborhood of v_0. The union $C^i(V)$, $i = 0, \ldots, q$, does fill out a full neighborhood of v_0.

If the cycle is accidental, for all $k = 1, \ldots n-1$, we have $v_k \neq v_0$, Note that, in this case, all the vertices of the cycle are distinct and $V = \bigcup_0^{n-1} W_k(F)$ is a full neighborhood of v_0.

The fundamental domain may have ideal vertices; that is, points on $\partial\Delta$ where two sides meet and are tangent. The finite equivalence classes of ideal vertices are called ideal cycles. If w_0 is an ideal vertex of an ideal cycle, the side labeling again gives us a shortest word for the cycle, $C = A_1 \ldots A_n$, that fixes w_0.

Because the translates of F fill out all of Δ, they must fill out a horocyclic neighborhood of a vertex w_0 of an ideal cycle. By Exercise 6.18 below, the sides of the translates meeting at an ideal vertex w_0 are all tangent; it takes infinitely many translates to fill out the whole horocyclic neighborhood. Thus the cycle has infinite order and the transformation that fixes w_0 is hyperbolic or parabolic. The cycle is called hyperbolic or parabolic accordingly.

In Example 2, the vertex of F at infinity is a cycle with one point and the transformation $B(z) = z + 4$ is the parabolic cycle word. In Example 3 the four ideal vertices belong to a single cycle which is parabolic (see Exercise 6.19 below) and, in Examples 4 and 5, the vertex at the origin is a cycle with one point and the transformation $A(z)$ fixing it is hyperbolic (see Exercise 6.20 below).

Now let θ_k be the interior angle of F between the two sides that meet at the vertex v_k of the cycle. Then θ_k is also the angle at $v_0 = W_k(v_k)$ of the translate $W_k(F)$. Since the union of the translates fills out a neighborhood or a horocycle of v_0 we have the *angle condition* for the cycle

$$\sum_{k=0}^{j-1} \theta_k = \frac{2\pi}{q}, \tag{6.2}$$

where $q = 1$ for accidental cycles and $q = \infty$ for ideal cycles.

Notice that cyclically permuting the word C determines the cycle transformations fixing the other vertices in the cycle.

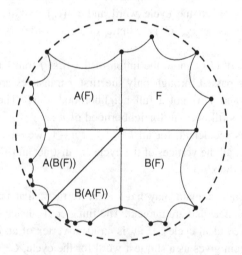

Figure 6.6   The vertex cycle at the origin

Remark. We deduce from the proof of Proposition 6.3.3 that, given a fundamental polygon for a Fuchsian group, we can read off a full presentation for the group: the side pairings are the generators and the accidental and elliptic cycle relations are the relators – words that determine the relations among the group elements. The ambiguity in the construction of a word from the side pairings comes from the relators. For example, if the path from F to $W(F)$ is perturbed through a vertex with relator $A_1 \ldots A_n$, the word changes by replacing part of the relator with an equivalent part, say $A_1 \ldots A_k$, $k < n$, with $A_n^{-1} \ldots A_{k+1}^{-1}$.

Let us find the cycles in the examples of the previous section. In Example 1, there is a vertex at $v_0 = (-1 + i\sqrt{3})/2$. Let s_1 be the vertical side $x = -1/2$; then s_1' is the side $x = 1/2$ and $v_1 = (1 + i\sqrt{3})/2$.

Next, s_2 is the side $|z| = 1$; it is paired with itself by $-1/z$ and the pairing sends v_1 back to v_0; thus the cycle consists of the two vertices $\{v_0, v_1\}$. The cycle transformation is $-1/(z-1)$ which is not the identity, but is an elliptic of order 3. The angles at the vertices are each $\pi/3$ and the sum of the angles of the cycle is $2\pi/3$.

The side s_2 contains the fixed point of $-1/z$ at i. We think of this as a vertex dividing s_2 and s_2' and the cycle contains just this one vertex. The cycle transformation for this vertex is just $-1/z$ and it is elliptic of order 2. There is a single angle of π in the cycle.

There is also an ideal vertex at infinity between the sides s_1 and s_1'. It is fixed by the parabolic cycle transformation $C = z + 1$ and the angle between the sides is zero. Let $D_2 = D \cap \{z = x + iy \,|\, y > 2\}$. The union $\bigcup_{-\infty}^{\infty} C^n(\overline{D_2})$ fills out the full horocycle at infinity, $\{y > 2\}$.

In Example 2, there is also a cycle consisting of the single ideal vertex at infinity between the sides s_2 and s_2'. It is fixed by the parabolic cycle transformation $z + 4$. Again, infinitely many translates by the cycle transformation fill out a horocycle at infinity.

The other end of s_2 is at -2, but only one side ends at this point and it is a free side. Its image under B is 2 and $B(\overline{F}) \cup \overline{F}$ fills out a neighborhood of the side s_2'. The same is true for the other endpoints. They are paired by A and $A(\overline{F}) \cup \overline{F}$ fills out a full neighborhood of s_1'.

In Example 4, there are two sides with one ideal and two free vertices. The ideal vertex is the fixed point of the hyperbolic transformation A. The translates of the polygon F fill out the right half of any horocycle at the origin.

All of the examples we have seen are for finitely presented Fuchsian groups; that is, groups that can be given by finitely many generators and finitely many relators. A theorem that is purely topological and follows directly from the definitions is

Theorem 6.4.1 *The fundamental group of a surface admits a finite presentation if and only if it is of finite topological type.*

It follows that a representation of the fundamental group of a surface as a Fuchsian group is finitely presented if and only if the surface is of finite topological type.

Surfaces that are not of finite topological type can also be represented as quotients of Fuchsian groups and these, by Theorem 6.4.1, have infinitely many generators. We sketch here a method of handling these surfaces and groups.

An exhaustion of a surface S is a sequence of subsurfaces

$$S_1 \subset S_2 \subset \cdots \subset S$$

such that S is the direct limit of the S_n. The fundamental groups satisfy

$$\pi_1(S_1) \subset \pi_2(S_2) \subset \cdots \subset \pi_1(S)$$

so that $\pi_1(S)$ is the direct limit of the $\pi_1(S_n)$. If S has infinite type, there exists an exhaustion by finite type surfaces, and one can represent their fundamental groups as an increasing sequence of finitely presented Fuchsian groups. Each of the approximating groups has a locally finite convex side-paired fundamental domain such that a fundamental domain for the limit group is the limit of the sequence of fundamental domains.

Exercise 6.16 Show that a vertex v_0 of a fundamental polygon for a group G is elliptic – that is, the fixed point of an elliptic element – if and only if every v_k in the cycle is elliptic.

Exercise 6.17 Show that if a cycle of vertices on the boundary of the hyperbolic plane has a parabolic cycle transformation C, with fixed point v, then the union of the translates of the fundamental polygon F by C^n, $n \in \mathbb{Z}$, fills out a horocycle at v.

Exercise 6.18 Show that a vertex of a fundamental polygon F cannot be in the same equivalence class as an ideal vertex. Show also that an ideal vertex and a free vertex cannot be in the same equivalence class.

Exercise 6.19 Show that the four vertices of the fundamental polygon of Example 3 are ideal vertices in the same equivalence class. Show that the cycle word is parabolic.

Exercise 6.20 Show that in Example 5 the vertex at the origin is an ideal vertex and the only one in its equivalence class. Show that the cycle word is hyperbolic.

Exercise 6.21 Show that a side of a fundamental polygon cannot be paired with more than one other side.

6.5 Poincaré's theorem

We saw in the previous section that, if we have a fundamental polygon for a group G that is locally finite and convex, then the side pairings generate the group and the angles at the vertices of a cycle satisfy the angle conditions (6.2).

In this section we prove Poincaré's polygon theorem which gives sufficient conditions for a polygon F with side pairings to be a fundamental polygon for the group G generated by the side pairings. In particular, the theorem says that the group generated by the side pairings is discontinuous and the only relations are those given by the cycles. The conditions include convexity and the angle condition we discussed above:

- (i) convexity,
- (ii) the angle conditions (6.2) on the cycles.

We need two more conditions to deduce that the group G is Fuchsian. These conditions will tell us not only that the group is Fuchsian but also that F is a locally finite fundamental polygon.

The first of these is a finiteness condition which is not surprising:

- (iii) A finiteness condition: under the projection $\pi_F : \overline{F} \to F^* = \overline{F}/\sim$, each point has only finitely many pre-images.

The density ρ projects to a density ρ_{F^*} on F^* via the projection π_F.

Definition 6.8 *Suppose a polygon F with side pairings satisfies conditions (i), (ii) and (iii). Define*

$$\rho_{F^*}(w_1, w_2) = \inf \int_\gamma \rho_{F^*}(t)dt$$

where by (iii) the infimum is realized by the projection of a piecewise geodesic joining $z_1, z_2 \in \overline{F}$ such that $\pi_F(z_i) = w_i, i = 1, 2$.

It is easy to see from the definition that ρ_{F^*} is symmetric and that, by condition (iii), $\rho_{F^*}(w_1, w_2) = 0$ if and only if w_1 and w_2 represent the same equivalence class. It also follows directly from the definition that the triangle inequality holds.

Note that when the group G is Fuchsian F^* can be identified with the quotient Δ/G and the metric ρ_{F^*} agrees with the metric Δ/G inherits from the universal cover $\pi : \Delta \to \Delta/G$.

To motivate the final condition we go back to our examples.

In Section 6.2 we saw that the half annulus $F = \{z \mid 1 < |z| < \lambda, \Im z > 0\}$ was a fundamental domain for the elementary group $G = \langle \lambda z \rangle$, considered as a Fuchsian group acting on $\mathbb{H}$. The quotient $\mathbb{H}/G$ is an open annulus and is identified with F^*.

Now consider the partial fundamental domain F of Example 4 of Section 6.3 for the same group with $\lambda = 3/2$. Recall that F was the domain in $\mathbb{H}$ bounded by the half circles $|z - 1/2| = 1/2$ and $|z - 3/4| = 3/4$. It had three vertices, one, a fixed point of the generator $z \mapsto \frac{3}{2}z$, and the others, the free vertices at 1 and 3/2. The quotient $F^* = \overline{F}/G$ is again topologically an open annulus but the metric ρ_{F^*} is not complete. To see this let γ be the segment of the geodesic $|z| = 1/2$ from its intersection with $\overline{F}$ at $(1 + \sqrt{3}i)/4$ to its intersection with the imaginary axis at $5i/4$.

Each of the infinitely many translates of F accumulating on the imaginary axis intersects γ. We can pull back the segment of γ in each of these translates to F using the group. We obtain a sequence of arcs $\sigma_i = A^{-i}(\gamma \cap A^i(\overline{F}))$ in

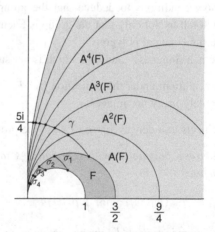

Figure 6.7   The geodesic γ has finite length

$\overline{F}$. Denote the ρ-length of σ_i by $l(\sigma_i)$. Then clearly, since γ has finite length, $\sum_0^\infty l(\sigma_i) < \infty$.

Denote the endpoints of σ_i on the lower and upper boundaries of F by z_i and z_i'. Now note that the points

$$w_i = \pi_F(z_i) = \pi_F(z_{i-1}')$$

converge to the boundary of $\overline{F}/G$. Given ϵ, however, we can choose m, n so large that

$$\rho_{F^*}(w_m, w_n) \le \sum_m^n \rho_{F^*}(w_i, w_{i-1}) \le \sum_{m-1}^\infty \rho(z_i, z_i') = \sum_{m-1}^\infty l(\sigma_i) < \epsilon$$

so that w_i is a Cauchy sequence on F^* with the induced metric ρ_{F^*} and the metric is not complete.

We can make similar examples for non-elementary groups.

In Example 2 of Section 6.3, the sides s_1 and s_1' are paired by the transformation A. Mimicking what we did in Example 4 of that section, we consider the geodesic l_1 with endpoints $(-\sqrt{3}, A(-\frac{1}{\sqrt{3}}))$. It is paired by A to the geodesic l_1' with endpoints $(A(\frac{1}{\sqrt{3}}), \sqrt{3})$. The polygon P_2 (see Figure 6.8) with sides l_1, l_1', s_2, s_2' is also a fundamental polygon for the group G generated by A and B. In fact, we can find a whole sequence of polygons P_{n+1} with sides l_n, l_n', s_2, s_2' where l_n is the geodesic with endpoints $(-\sqrt{3}, A^n(-\frac{1}{\sqrt{3}}))$ and l_n'

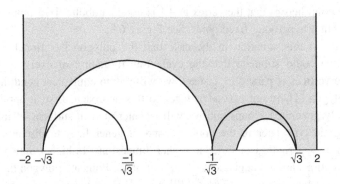

Figure 6.8 The polygon P_2

is its image under A, the geodesic with endpoint $(A^n(\frac{1}{\sqrt{3}}), \sqrt{3})$. If we let n tend to infinity, the polygons P_n tend to a limit polygon P_∞. The sides l_n, l'_n tend to sides l_∞, l'_∞ with common endpoint $+1$, the attracting fixed point of A. The polygon P_∞ is convex, satisfies the finiteness condition, and satisfies the angle condition for the vertex at infinity.

The polygon P_∞ satisfies the first condition for a fundamental domain for G. If any two points in P_∞ were equivalent under the group, they would be equivalent in some P_n, contradicting the fact that the P_n are all fundamental polygons.

We have to see if it satisfies the second condition. Consider the vertex at $+1$. The two sides that meet there make a zero angle so satisfy the angle condition. Now let us look at translates of P_∞.

The image $A(P_\infty)$ is a polygon lying under l'_∞ sharing the vertex $+1$. The images $A^n(P_\infty)$ contain sides $A^n(l_\infty)$ that accumulate to $+1$. Now consider the images $A^{-n}(P_\infty)$ that contain sides $A^{-n}(l_\infty)$. These sides accumulate onto the axis Ax_A joining $-1, +1$ and the images all lie above this axis. The transformations $A^{\pm n}$ map the half plane R under Ax_A to itself and the half plane R' outside Ax_A to itself. The transformations $B^{\pm n}$ map R into R'. Checking carefully, it is not to hard to see that no element of G can map P_∞ inside R. In fact, if we conjugate the group so that the axis of A becomes the imaginary axis, the picture near the axis is exactly like the picture for the elementary group of Example 4. Therefore

$$\bigcup_{W \in G} W(\overline{P_\infty}) = \mathbb{H} \setminus \bigcup_{W \in G} R.$$

That is, the translates of P_∞ cover a domain in $\mathbb{H}$ that is missing a countable number of half planes so it does not satisfy the second condition.

The point here is that the vertex at $+1$ is not a parabolic fixed point of the group but a hyperbolic fixed point. See Figure 6.9.

To avoid this situation in the case that the polygon has finitely many sides, we could stipulate that the cycle transformation of every cycle of infinite vertices is parabolic. Instead, we will put an equivalent condition on the polygon. There are two advantages to this approach: first, it applies to infinitely generated groups and we will be interested in surfaces of infinite topological type later in the book; second, it generalizes to discontinuous groups of hyperbolic isometries in higher dimensions. In higher dimensions, the conditions are not equivalent. The proof of the Poincaré polygon theorem we present here is due to Maskit, [39], sec. IV.H. Poincaré's original proof and many others that followed it had gaps.

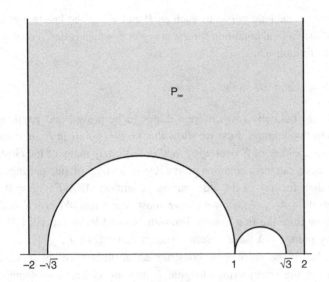

Figure 6.9   The polygon P_∞

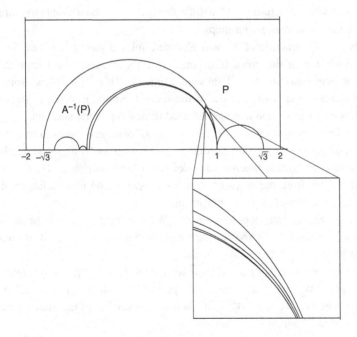

Figure 6.10   The blow-up exhibits the non-local-finiteness of the fundamental domain $P = P_\infty$

In order to exclude polygons such as P and P_∞ in the last two examples, we introduce a final condition for the polygon F which is based on the metric ρ_{F^*} of Definition 6.8.

- (iv) F^* is complete in the ρ_{F^*}-metric.

There are basically two different things to be proved and we divide the proof into two lemmas. First we show that no two points in F are conjugate; that is, no translate of F overlaps F and only finitely many of the closures of the translates can have common points. Definition 6.5 of side pairings makes it clear that, for any of the side pairing generators A, $A(F) \cap F = \emptyset$. If we consider $W = A_1 \ldots A_n$, however, we must prove that $W(F) \cap F = \emptyset$. This will follow from the first lemma. From the second lemma it will follow that for every point $z \in \Delta$ there is some W such that $W(z) \in \overline{F}$.

The idea of the proof is to construct an abstract covering space Δ^* for Δ by using the side pairings to glue $\overline{F}$ and all its translates together. By construction there will be no overlaps in Δ^*; any overlap of $W(F)$ and F in Δ will lift to different sheets of Δ^*. This is similar to the construction of the universal covering space in Chapter 4. This is the content of the first lemma. The conformal structure on Δ^* will be defined by the conformal structure on F and the conformal gluing maps.

The metric structure of F^* will also determine a metric structure for Δ^*. The final step in the proof is to show that the covering map from Δ^* to Δ is a homeomorphism and thus an isometry. This will follow from the last condition and is the content of the second lemma. In order to state this condition and prove the lemmas we need to develop some notation.

The first step is to form an abstract group G^* with generators in one to one correspondence with the side pairings and relators given by the cycle relations. The generator correspondence thus defines a homomorphism $H: G^* \to G$. It will follow from the second lemma that there are no other relations in G and therefore that H is an isomorphism.

We put the discrete topology on G^*. We use the notation A for the side pairings in G and A^* for the generators of G^*. Similarly, we use W for words in the generators of G and W^* for words in G^*.

Consider the space of pairs $G^* \times \overline{F}$ with the following equivalence relation: $(W_1^*, z_1) \sim (W_2^*, z_2)$ if there is a side pairing A with $A(z_1) = z_2$ and if, as elements of G^*, $W_2^* = W_1^* A^{-1}$. This is a formal version of the gluing process described in Theorem 4.4.1. Set

$$\Delta^* = (G^* \times \overline{F})/\sim .$$

We use the product here because we will want the hyperbolic structure on the second factor induced by restriction. Use the usual identification topology on the quotient so that the projection $\pi: G^* \times \overline{F} \to \Delta^*$ is continuous.

Note that we do not yet know that there are only finitely many points in $G^* \times \overline{F}$ projecting to the same point in Δ^*. This will be a consequence of Lemma 6.5.1.

We have two maps:

$$g: \Delta^* \to F^*$$

defined by first projecting onto the second factor and then applying the map π_F and

$$h: \Delta^* \to \Delta$$

defined by $h(W^*, z) = H(W^*)(z)$. Since the projection π is continuous, it follows easily that h is continuous. The pullback of this map defines the structure on Δ^*.

Thus Δ^* can be thought of in two ways: first as the set of translates of $\overline{F}$ under G^* glued together along the boundaries in such a way that the map h is well defined and second as the set of translates of $\overline{F}$ under G sewn together so that any non-trivial overlap occurs on different sheets. The map h projects this covering back to Δ.

Now we prove that h is a local homeomorphism.

Lemma 6.5.1 *Let F be a polygon with side pairings satisfying conditions (i)–(iii). Then $h: \Delta^* \to \Delta$ is a regular covering.*

Proof. The group G^* acts on the space Δ^* as a group of homeomorphisms. A fundamental domain for the action is the set $D = (Id \times F)$. By construction, as W^* varies in G^*, the translates $W^*(D)$ are disjoint and the translates cover Δ^*.

If $z \in \Delta$ is in the image of h, there is a $W \in G$ such that $W(z) \in \overline{F}$ and $z^* = \pi_F(W(z)) \in F^*$. Therefore, the lemma will follow if we prove that every point $z^* \in F^*$ has a neighborhood U such that $g^{-1}(U)$ is a disjoint union of sets U_k, whose closures are relatively compact in Δ^*, and, in addition, $h|_{U_k}$ is a homeomorphism onto a disk in Δ.

Let z be a point of $\overline{F}$. We consider three cases. First, suppose it is an interior point of F. We can find a disk V of radius δ centered at z such that $V \subset F$. Then $g^{-1}(V)$ is a disjoint union $\bigoplus U_k$ of sets of the form $(W^* \times V)$. If we set $U = \pi_F(V)$, then $h|_{U_k}: U_k \to U$ is a homeomorphism onto a disk.

Now suppose z is a point on the interior of a side s. Let A be the side pairing so that $z' = A(z)$ and $s' = A(s)$. We want to find a disk neighborhood of $z^* = \pi_F(z)$. We can find $\delta > 0$ so that the half disks V, V' of radius $\delta/2$, centered at z, z' and with diameter on s, s' respectively, bound disjoint domains with interiors completely contained in F. We assume these half disks contain the diameter. Each connected component U_k of $g^{-1}(V \cup V')$ is a disk in Δ^*. For example, one such component containing (Id, z) is the disk $(Id \times V) \cup (A^{-1} \times V')$. Set $U = \pi_F(V \cup V')$; then $h|_{U_k} : U_k \to U$ is a homeomorphism onto a disk.

Finally suppose that z is a vertex. Let $z = z_0$ and let $z_0, z_1, \ldots, z_{n-1}$ be the vertices in its cycle. Let the cycle transformation be $C = A_n \ldots A_1$ and suppose it satisfies the relation $C^q = Id$. Rewrite this relation as

$$A_1^{-1} \ldots A_n^{-1} A_1^{-1} \ldots A_n^{-1} \ldots A_1^{-1} \ldots A_n^{-1} = Id$$

and let $W_k = A_1^{-1} \ldots A_n^{-1} \ldots A_1^{-1} \ldots A_{n_k}^{-1}$ be the product of the first k generators appearing in the relation.

For $i = 0, \ldots, n-1$, we can find a set of disjoint sectors V_i with vertex z_i, radius δ and sides the pair of sides of F that meet at z_i, whose interiors are completely contained in F. Set $W_i(V_i) = V_i'$. Next set $V = \bigcup_0^{n-1} V_i'$. Then, by the angle condition, each component U_k of $g^{-1}(\bigcup_0^{q-1} C^j(V))$ is a disk in Δ^*. For example, one such component containing (Id, z_0) is the disk $\bigcup_0^{q-1} (C^j \times V)$. Set $U = \pi_F(\bigcup_0^{q-1} C^j(V))$; then $h|_{U_k} : U_k \to U$ is a homeomorphism onto a disk. $\square$

It follows from the proof of the lemma that only finitely many translates of $(Id \times \overline{F})$ meet at any point so that the tessellation by translates satisfies the finiteness condition.

Next we prove that h is a homeomorphism provided F satisfies condition (iv) in addition to the others.

Lemma 6.5.2 *Let F be a polygon with side pairings satisfying conditions (i)–(iv). Then $h : \Delta^* \to \Delta$ is a homeomorphism.*

Proof. We need only show that any infinite geodesic in Δ can be lifted and hence that h is onto. Since Δ is simply connected it follows that the covering h is a homeomorphism.

Let z be a point in $h(\Delta^*)$. Let $\gamma(t)$ be a geodesic emanating from z parametrized by arclength. Since h is a covering, we can take the local lift $\tilde{\gamma}$ of γ to Δ^* at the point (Id, z). If we cannot extend the lift along the whole geodesic there will be some $t_0 < \infty$ such that we can continue the lift only for $0 < t < t_0$, but not beyond. We can find a sequence t_n of real numbers such that, for each n, the segment of $\tilde{\gamma}(t)$ between $\tilde{\gamma}(t_n)$ and $\tilde{\gamma}(t_{n+1})$ is in some disk U_n on which h is a homeomorphism. We can assume that, along each of these segments, the first factor W_n is constant and by the definition

$$\rho_{F^*}(g(\tilde{\gamma}(t_n)), g(\tilde{\gamma}(t_{n+1}))) \leq \rho_\Delta(\gamma(t_n), \gamma(t_{n+1})).$$

We then have

$$\sum \rho_{F^*}(g(\tilde{\gamma}(t_n)), g(\tilde{\gamma}(t_{n+1}))) \leq \sum \rho_\Delta(\gamma(t_n), \gamma(t_{n+1})) < \infty.$$

Since ρ_{F^*} is complete, we can find a point $(W^*, z_0) \in \Delta^*$ such that $\tilde{\gamma}(t_n) \to$ (W^*, z_0). Therefore, for n sufficiently large, $\gamma(t_n)$ lies in a neighborhood U of $h(W^*, z_0)$ on which h has a well defined local inverse T. It follows that, as $t \to t_0$, $\tilde{\gamma}(t)$ can be continued to $T(U)$ and hence that the whole geodesic can be lifted. $\square$

We summarize these results as

Theorem 6.5.1 (Poincaré polygon theorem) *Let F be a hyperbolic polygon in Δ (or $\mathbb{H}$) with side pairings. If F satisfies conditions (i), (ii), (iii) and (iv), then the group G generated by the side pairings is discrete and F is a fundamental polygon for G*

Exercise 6.22 Look at Example 5 above where there is a hyperbolic vertex and show that ρ_{F^*} is not complete.

Exercise 6.23 Let G be a discontinuous group generated by the map $z \mapsto 3z$. Show that $\rho_{\Delta/G}(w_1, w_2) = \inf \rho(z_1, z_2)$ where the infimum is over all points $z_1, z_2 \in \Delta$ such that the G-equivalence class of z_1 is w_1 and the G-equivalence class of z_2 is w_2 defines another metric on Δ/G. Show that $\rho_{\Delta/G}$ is complete. Evaluate $\rho_{\Delta/G}([\frac{1}{2}], [\frac{i}{2}])$ and $\rho_{F^*}([\frac{1}{2}], [\frac{i}{2}])$ where $F = \{z | 1 < |z| < 3\}$.

7

The hyperbolic metric for arbitrary domains

7.1 Definition of the hyperbolic metric

In Chapter 6 the metric ρ_{F_*} on F^* was inherited from its covering space. Similarly, in Exercise 6.23 the metric on the annulus comes from the metric on the covering space. In this chapter we generalize this method to define a metric on an arbitrary Riemann surface from the metric on its covering space. Since our primary interest in the following chapters will be in plane domains we will restrict our discussion to Riemann surfaces that are plane domains. For readability, we tacitly assume all universal covering maps are holomorphic.

Definition 7.1 *A plane domain X with at least two boundary points is called a hyperbolic domain.*

By the uniformization theorem, Theorem 4.6.1, there exists a universal covering map π from the unit disk Δ to any hyperbolic domain X. We push the hyperbolic density in the unit disk defined in Chapter 2 by the map π into a density defined on X as follows.

Definition 7.2 *If z is any point in X and t is any point in Δ with $\pi(t) = z$, then the hyperbolic density on X at the point z is defined by*

$$\rho_X(z) = \frac{\rho(t)}{|\pi'(t)|}. \tag{7.1}$$

To see that the hyperbolic density is well defined, suppose that t and s are two points in the unit disk such that $\pi(t) = \pi(s) = z$; that is, t and s are two pre-images of z under the covering map π. Then by Theorem 4.5.1 there exists a homeomorphism f from Δ onto Δ such that $f(t) = s$ and $\pi \circ f = \pi$. Since π is holomorphic, f is also. Therefore, by Theorem 3.2.2,

$$\frac{\rho(t)}{|\pi'(t)|} = \frac{\rho(t)}{|(\pi \circ f)'(t)|} = \frac{\rho(t)}{|f'(t)|} \frac{1}{|\pi'(s)|} = \frac{\rho(s)}{|\pi'(s)|}.$$

Now that we have a hyperbolic density defined on X, we make it into a hyperbolic metric on X: we integrate the density $\rho_X(t)|dt|$ to measure the hyperbolic distance between any two points on the domain X. For any rectifiable path γ, we set

$$\rho_X(\gamma) = \int_\gamma \rho_X(t)|dt|.$$

Then we define

Definition 7.3

$$\rho_X(p, q) = \inf \rho_X(\gamma),$$

where the infimum is over all paths γ in X joining p to q.

The relation (7.1) says that a universal covering map is an infinitesimal hyperbolic isometry. The same is true for any holomorphic covering map.

Theorem 7.1.1 *Suppose that g is a regular holomorphic covering map from a hyperbolic plane domain Ω onto a plane domain X. Then g is an infinitesimal isometry. That is,*

$$\rho_X(g(t))|g'(t)| = \rho_\Omega(t)$$

for all t in Ω.

Proof. If π is a universal covering map from the unit disk onto Ω, then the composition $g \circ \pi$ is a universal covering map from Δ onto X. Indeed, if γ is any curve in X it can be lifted to Δ first by lifting it to Ω using any inverse branch of g and then by lifting the image in Ω to Δ by any inverse branch of π. Now for any pre-images $t = g^{-1}(z)$ and $s = \pi^{-1}(t)$ the relation (7.1) yields

$$\rho_\Omega(t)|\pi'(s)| = \rho(s) = \rho_X(g(t))|(g \circ \pi)'(s)|.$$

Therefore, by the chain rule,

$$\rho_\Omega(t)\,|\pi'(s)| = \rho_X(g(t))|g'(t)||\pi'(s)|,$$

so that

$$\rho_\Omega(t) = \rho_X(g(t))|g'(t)|. \qquad \square$$

It follows that any regular holomorphic covering surjection g between two plane domains Ω and X preserves the hyperbolic lengths of curves: $\rho_\Omega(\gamma) = \rho_X(g(\gamma))$. We deduce from Definition 7.3 that the hyperbolic distance between any two points z and w in X is equal to the infimum of the hyperbolic distances between the set of pre-images of z and the set of pre-images of w in Ω. That is,

$$\rho_X(z, w) = \inf\{\rho_\Omega(t, s) \mid t, s \in \Omega, \ g(t) = z, g(s) = w\}. \qquad (7.2)$$

If we apply this equality to the universal covering map π from the unit disk to any plane domain X we obtain the following description of the hyperbolic distance on any plane domain:

$$\rho_X(z, w) = \inf\{\rho(t, s) \mid t, s \in \Delta, \ \pi(t) = z, \ \pi(s) = w\}.$$

Since π is a continuous map, we can say more.

Theorem 7.1.2 *Let π be a universal covering map from the unit disk onto a plane domain X. If z and w are any two points in X, and t is any pre-image of z in Δ, then*

$$\rho_X(z, w) = \min\{\rho(t, s) \mid s \in \Delta, \ \pi(s) = w\}.$$

Let us now verify that ρ_X is indeed a metric on X.

Theorem 7.1.3 *For every hyperbolic plane domain X, (X, ρ_X) is a complete metric space.*

Proof. Definition 7.3 implies that $\rho_X(z, w)$ is symmetric and satisfies the triangle inequality. Let z and w be any two distinct points in X. Let t be a pre-image of z by a universal covering map π. Theorem 7.1.2 implies that there exists a point s in Δ such that $\pi(s) = w$ and $\rho_X(z, w) = \rho(t, s)$. Since t and s have different images under π, they must also be distinct. Since $\rho(t, s)$ is non-negative and zero only if $t = s$, we conclude that $\rho_X(z, w) = \rho(t, s) > 0$. Therefore, (X, ρ_X) is a metric space.

To prove completeness, suppose that p_n is a Cauchy sequence in (X, ρ_X). Take any pre-image t_1 of p_1 under the universal covering map π. By Theorem 7.1.2 there are points t_n in Δ such that, for all positive integers n, $\pi(t_n) = p_n$ and $\rho(t_1, t_n) = \rho_X(p_1, p_n)$. Therefore t_n is a bounded sequence in the closed unit disk so there exists a subsequence t_{n_k} which converges in the Euclidean metric to a point t in the closed disk. Since $\rho(t_1, t_n)$ is bounded, t belongs to Δ and thus, by formula (2.4), t_{n_k} converges to t in the hyperbolic metric ρ.

Since π is continuous, $\rho_X(\pi(t), p_{n_k}) \to 0$ and the Cauchy sequence p_n has a convergent subsequence converging to $\pi(t)$; thus, the whole sequence converges in the ρ_X metric to the point $\pi(t)$. $\quad\square$

Exercise 7.1 Let G be the subgroup of $PSL(2, \mathbb{R})$ acting on $\mathbb{H}$ generated by $A(z) = 6z$. Verify that $F = \{z \mid 1 < |z| < 6\} \cap \mathbb{H}$ is a fundamental domain for G. Verify that $X = \mathbb{H}/G$ is homeomorphic to $S = \{z \mid 1 \leq |z| < 6\} \cap \mathbb{H}$ (with respect to the quotient topology) and that both metrics ρ_X and ρ_{F*} carry over to S. Compare these metrics and evaluate the distance from i to $5i$ with respect to both metrics.

7.2 Properties of the hyperbolic metric for X

Next we study some of the basic properties of the hyperbolic metric on a hyperbolic domain X.

Theorem 7.2.1 *The hyperbolic density on a hyperbolic domain X is an infinitesimal form of the hyperbolic distance. That is, if z is a point in X then*

$$\rho_X(z, z+t) = |t|\rho_X(z) + o(t).$$

Proof. Take any pre-image a of z under the universal covering map π from Δ onto X. By Theorem 7.1.2, there exists a point a_t in Δ, such that $\pi(a_t) = z + t$ and $\rho(a, a_t) = \rho_X(z, z+t)$. Furthermore, as $t \to 0$, we have $a_t \to a$. Thus,

$$\frac{\rho_X(z, z+t)}{|t|} = \frac{\rho(a, a_t)}{|t|} = \frac{\rho(a, a_t)}{|a_t - a|}\left|\frac{a_t - a}{t}\right|.$$

The equality (2.8) implies

$$\frac{\rho(a, a_t)}{|a_t - a|} \to \rho(a), \quad \text{as } t \to 0.$$

Furthermore,

$$\left|\frac{a_t - a}{t}\right| = \left|\frac{a_t - a}{(z+t) - z}\right| = \left|\frac{a_t - a}{\pi(a_t) - \pi(a)}\right|$$

so that

$$\left|\frac{a_t - a}{t}\right| \to \frac{1}{|\pi'(a)|} \quad \text{as } t \to 0.$$

Therefore, by the definition of the density, formula (7.1),

$$\frac{\rho_X(z, z+t)}{|t|} \to \rho(a)\frac{1}{|\pi'(a)|} = \rho_X(\pi(a)) \text{ as } t \to 0. \quad \square$$

Since the Euclidean metric d_X on X satisfies the formula $d_X(z, z+t) = |t|d_X(z)$ with $d_X(z) \equiv 1$, we have the following corollary to Theorem 7.2.1.

Corollary 7.2.1 *The hyperbolic metric on any hyperbolic domain is locally equivalent to the Euclidean metric.*

Theorem 7.2.2 *The hyperbolic density function $\rho_X(z)$ is a positive continuous function.*

Proof. Let z_0 be any point in X. Pick any pre-image t_0 of z_0 under the universal covering map π from Δ onto X. Since π is holomorphic and locally one-to-one, there exists a local inverse g of π defined in a neighborhood N of z_0. If $z \in N$ and $t = g(z)$ the formulas (7.1) and (2.2) yield

$$\rho_X(z) = \frac{\rho(t)}{|\pi'(t)|} = \rho(g(z))|g'(z)| = \frac{|g'(z)|}{1 - |g(z)|^2} > 0.$$

Thus, $z \mapsto \rho_X(z)$ is a positive continuous function. $\square$

In Chapter 2, we defined a hyperbolic geodesic in Δ as a curve on which the triangle inequality is an equality for any three points in order. We use the covering projection to define geodesics on an arbitrary domain X.

Definition 7.4 *Let X be a hyperbolic domain and let $\pi : \Delta \to X$ be a universal covering map. A curve γ on X is a geodesic on X if and only if every lift $\pi^{-1}(\gamma)$ is a geodesic in Δ.*

A local version of Definition 2.2 together with Theorems 2.1.1 and 7.1.1 implies

Proposition 7.2.1 *If γ is a curve in the hyperbolic domain X such that for every triple of points p, r, q on γ with r between p and q we have*

$$\rho_X(p, q) = \rho_X(p, r) + \rho_X(r, q),$$

then γ is a geodesic *on X.*

As for the unit disk, we have

Theorem 7.2.3 *For any two distinct points z and w in the domain X, there exists at least one shortest path γ joining z to w. Furthermore γ is a geodesic.*

Proof. Let z and w be any two distinct points in X. By Theorem 7.1.2 there exist two points t and s in the unit disk such that $\pi(t) = z$, $\pi(s) = w$ and $\rho(t, s) = \rho_X(z, w)$. By Theorem 2.1.1 there exists a geodesic γ in Δ which joins t and s. By Definition 7.4, the curve $\pi(\gamma)$ is then a curve that joins

z and w and by Theorem 4.5.1 is a geodesic on X. Because the covering map π preserves the lengths of curves we have

$$\rho_X(z, w) = \rho(t, s) = \rho(\gamma) = \rho_X(\pi(\gamma)). \quad \square$$

Note, however, that this proof does not show that $\pi(\gamma)$ is the only geodesic which joins z and w, but it shows it is the shortest one. The following theorem says that if we think of the curve joining z and w as a piece of hyperbolic elastic stretched along the curve, and if we only hold it down at the ends, the elastic will pull tight to a geodesic joining the endpoints.

Theorem 7.2.4 *Let z and w be any two distinct points on X. Then there is a unique geodesic in every homotopy class of curves joining z and w.*

Proof. Let z, w be distinct points in X and let δ be any curve in a given homotopy class of curves joining z and w. Let $s \in \Delta$ be any lift of z and let $\tilde{\delta}$ be a lift of δ at s. Let t be the endpoint of $\tilde{\delta}$. By the monodromy theorem, t is uniquely determined by the homotopy class of δ.

There is a unique geodesic $\tilde{\gamma}$ in Δ joining s and t. The projection $\gamma = \pi(\tilde{\gamma})$ is a geodesic on X in the homotopy class of δ.

To see that γ is unique, suppose γ and γ' are homotopic geodesics joining z to w on X. By the monodromy theorem, the lifts $\tilde{\gamma}$ and $\tilde{\gamma}'$ at s are both geodesic and end at t and thus are the same. $\quad \square$

The above theorem holds for convex Euclidean as well as hyperbolic surfaces. In the next theorem we consider closed curves made of elastic bands. If we stretch an elastic band along a non-homotopically-trivial closed curve and then let go, it will pull tight on the surface. On a Euclidean surface such as a torus, the tight curve can be slid along and is not unique, but, on a hyperbolic surface, it cannot slide – it is unique, as the next theorem tells us.

Theorem 7.2.5 *Let γ be a non-trivial closed curve on X. Then either there is a unique shortest geodesic in the free homotopy class of γ, or γ is homotopic to a puncture on X and there are arbitrarily short curves in its free homotopy class.*

Proof. Let G be a Fuchsian group such that $X = \Delta/G$. Choose a point z on γ and a point $s \in \Delta$ such that $\pi(s) = z$. Let $\tilde{\gamma}$ be the lift of γ to s and let s' be the endpoint of $\tilde{\gamma}$. Then, by Theorem 4.5.1, there is a unique element $A \in G$ such that $A(s) = s'$. Since γ is not homotopically trivial, $A \neq Id$.

Since G is a representation of $\pi_1(X)$ it has no elliptic elements and A is either parabolic or hyperbolic. If it is hyperbolic, let Ax_A be the axis of A and let p be any point on Ax_A. Recall that, by Definition 2.5, Ax_A is the geodesic joining the fixed points of A and so is uniquely determined by A. Let $\tilde{\sigma}$ be the segment of Ax_A between p and $A(p)$. Join s to p in Δ by a curve α. Then $\alpha' = A(\alpha)$ is a curve in Δ joining s' to $A(p)$ and $\alpha'^{-1}\tilde{\gamma}\alpha'$ is homotopic to $\tilde{\sigma}$ in Δ. It follows that the projections $\pi(\alpha^{-1}\tilde{\gamma}\alpha')$ and $\sigma = \pi(\tilde{\sigma})$ are homotopic on X, so that, by Definition 4.8, σ is freely homotopic to γ.

Let us see how the construction of σ depends on the choice of lift. Suppose t is a different lift of z and $\tilde{\gamma}'$ is the lift of γ at t. Then, by Theorem 4.5.1, there is a unique element $B \in G$ such that $B(s) = t$. The terminal endpoint of $\tilde{\gamma}'$ is $BAB^{-1}(t)$. Under the covering projection, by the group invariance, $\pi(Ax_{BAB^{-1}}) = \pi(Ax_A) = \sigma$ so σ is uniquely determined.

Next suppose A is parabolic. Pick $t \in \Delta$, let $t' = A(t)$ and let $\tilde{\sigma}$ be a geodesic joining t and t'. As we did above, we can find α and $\alpha' = A(\alpha)$ so that $\alpha^{-1}\tilde{\gamma}\alpha'$ is homotopic to $\tilde{\sigma}$ in Δ and $\sigma = \pi(\tilde{\sigma})$ is freely homotopic to γ on X. By formula (2.15), we can choose $\tilde{\sigma}$ to have length as small as we like. $\square$

Exercise 7.2 Show that all the lifts of a closed geodesic that is shortest in its free homotopy class are axes of elements in a single conjugacy class.

7.3 The Schwarz–Pick lemma

The very useful Schwarz–Pick lemma has a generalization to arbitrary domains.

Theorem 7.3.1 (Schwarz lemma for arbitrary domains) *If f is a holomorphic map from a domain Ω into a domain X, then f is both an infinitesimal and a global contraction with respect to the corresponding hyperbolic metrics on Ω and X. That is,*

$$\rho_X(f(t))|f'(t)| \leq \rho_\Omega(t) \text{ for all } t \in \Omega \tag{7.3}$$

and

$$\rho_X(f(t), f(s)) \leq \rho_\Omega(t, s) \text{ for all } t, s \in \Omega. \tag{7.4}$$

Proof. Let π_Ω and π_X be the universal covering maps from the unit disk to Ω and X respectively. Pick pre-images p and q in Δ of t in Ω and $f(t)$ in X under π_Ω and π_X respectively. Now lift the map f to a map $\tilde{f}$ from Δ into Δ as follows. Pick an arbitrary point a in Δ. Join the point a to the point p by any curve γ. Then project the curve γ down to the curve $\pi(\gamma)$ in Ω and take its image $f(\pi(\gamma))$ in X. Finally lift the curve $f(\pi(\gamma))$ to a curve $\tilde{\gamma}$ in Δ that starts at the point q. The endpoint of $\tilde{\gamma}$ is by definition the image of a. By the monodromy theorem, the resulting map $\tilde{f}$ is well defined because the unit disk Δ is simply connected. Since the covering maps are holomorphic and locally one to one, the map $\tilde{f}$ is holomorphic. Furthermore, we have

$$f \circ \pi_\Omega(a) = \pi_X \circ \tilde{f}(a)$$

for all a in Δ. Therefore, taking derivatives on both sides we obtain

$$f'(\pi_\Omega(a))\pi'_\Omega(a) = \pi'_X(\tilde{f}(a))\tilde{f}'(a). \tag{7.5}$$

Applying the Schwarz–Pick lemma, Theorem 3.2.2, we have

$$\rho(\tilde{f}(a))|\tilde{f}'(a)| \le \rho(a). \tag{7.6}$$

Combining (7.5), (7.6) and (7.1) we obtain

$$\rho(\tilde{f}(a))|f'(\pi_\Omega(a))\pi'_\Omega(a)| \le \rho(a)|\pi'_X(\tilde{f}(a))|,$$

$$\rho_X(\pi_X(\tilde{f}(a)))|f'(\pi_\Omega(a))| \le \rho_\Omega(\pi_\Omega(a)),$$

$$\rho_X(f(\pi_\Omega(a)))|f'(\pi_\Omega(a))| \le \rho_\Omega(\pi_\Omega(a)).$$

Since π_Ω is surjective, (7.3) holds.

To show (7.4), let s and t be any two points in Ω. By Theorem 7.2.3, there exists a shortest path γ in Ω joining s and t. Using formula (7.3), we obtain

$$\rho_X(f(s), f(t)) \le \rho_X(f(\gamma)) = \int_{f(\gamma)} \rho_X(\tau)|d\tau| = \int_\gamma \rho_X(f(\sigma))|f'(\sigma)||d\sigma|$$

$$\le \int_\gamma \rho_\Omega(\sigma)|d\sigma| = \rho_\Omega(\gamma) = \rho_\Omega(s, t). \quad \square$$

If f is a conformal homeomorphism, then we may apply Theorem 7.3.1 to both f and its holomorphic inverse f^{-1}. Thus, we obtain the following corollary of Theorem 7.3.1.

Corollary 7.3.1 *If f is a conformal homeomorphism from a domain Ω onto a domain X, then f is both an infinitesimal and a global isometry with respect to the corresponding hyperbolic metrics on Ω and X. That is,*

$$\rho_X(f(z))|f'(z)| = \rho_\Omega(z) \text{ for all } z \in \Omega \tag{7.7}$$

and

$$\rho_X(f(z), f(w)) = \rho_\Omega(z, w) \text{ for all } z, w \in \Omega. \tag{7.8}$$

Theorem 7.3.1 is a generalization of Theorem 3.2.2 to arbitrary plane domains. If Ω is a universal cover of a non-simply-connected domain X, the universal covering map π_X is surjective and a local homeomorphism. In this case, Definitions 7.2 and 7.3 tell us that we have equality in formula (7.3) and local equality in formula (7.4). Because π_X is not injective, it has no inverse so that Theorem 3.2.3 does not generalize completely.

The next corollary provides another approach to hyperbolic density, via holomorphic mappings.

Corollary 7.3.2 *If X is a hyperbolic plane domain and z and w are any two points in X, then*

$$\rho_X(z) = \inf \frac{1}{|f'(0)|} \tag{7.9}$$

where the infimum is over all holomorphic maps f from the unit disk to X such that $f(0) = z$. Alternatively,

$$\rho_X(z) = \inf \frac{\rho(t)}{|f'(t)|} \tag{7.10}$$

where the infimum is over all holomorphic maps f from the unit disk to X and all t in Δ, such that $f(t) = z$.

In addition, on the global level,

$$\rho_X(z, w) = \inf \rho(0, s) \tag{7.11}$$

where the infimum is over all holomorphic maps f from the unit disk to X and all s in Δ, such that $f(0) = z$ and $f(s) = w$. Alternatively,

$$\rho_X(z, w) = \inf \rho(t, s) \tag{7.12}$$

where the infimum is over all holomorphic maps f from the unit disk to X and all t and s in Δ, such that $f(t) = z$ and $f(s) = w$.

Proof. Formulas (7.9) and (7.10) follow from formulas (7.1) and (7.3) by observing that any universal covering map from Δ onto X may be precomposed with a Möbius transformation. Similarly, formulas (7.11) and (7.12) follow from inequality (7.4) and Theorem 7.1.2. $\square$

Exercise 7.3 Let X be any hyperbolic plane domain. Define the function $f(z, w) = \rho_X(z, w)$. At what pairs of points (z, w) is f continuous?

Exercise 7.4 Let Ω be any hyperbolic plane domain, and let z be a point in Ω. Show that, if t_n is a sequence of points in Ω that converges (in the Euclidean metric) to a point t outside Ω, then $\rho_\Omega(z, t_n) \to \infty$. Hint: Take a holomorphic universal covering map π from the unit disk onto Ω, such that $\pi(0) = z$. If D is a hyperbolic disk in Δ with center at 0 and hyperbolic radius C, $\bar{D}$ is compact, and so is $\pi(\bar{D})$.

Exercise 7.5 Let a be a point in a subdomain X of Δ. Show that, if a is deep enough in X, then the ρ_X and ρ metrics are almost the same. Precisely, let $C = R(X, \Delta, a)$ be the radius of the largest ρ-disk with center at a which is inside X and assume $C > 1$. Show that if z is a point in X such that $\rho(a, z) < 1$ then

$$\rho_X(a, z) \leq (1 + \epsilon)\rho(a, z)$$

where $\epsilon = \epsilon(C) \to 0$ as $C \to \infty$.

7.4 Examples

Now is the time to look at some examples.

Example 1 X is simply connected.

If X is simply connected, then, by the Riemann mapping theorem, there exists a conformal homeomorphism f from the unit disk Δ onto X and, by Definition 7.2, formula (7.1), or Corollary 7.3.1,

$$\rho_X(f(t)) = \frac{\rho(t)}{|f'(t)|}. \tag{7.13}$$

It is not always easy, however, to find the Riemann mapping for a given simply connected domain. Below we give three examples.

The first example of a plane domain for which there is an easy Riemann map is the upper half plane $\mathbb{H}$. In this case, up to pre-composition with a Möbius transformation leaving Δ invariant, the Riemann map $f_{\mathbb{H}} : \Delta \to \mathbb{H}$ is $f_{\mathbb{H}}(z) = i\frac{1+z}{1-z}$ and the inverse h of $f_{\mathbb{H}}$ is $h(w) = \frac{w-i}{w+i}$. Therefore,

$$\rho_{\mathbb{H}}(w) = \rho(h(w))|h'(w)| = \frac{|h'(w)|}{1 - |h(w)|^2},$$

$$\rho_{\mathbb{H}}(w) = \frac{2}{|w + i|^2 - |w - i|^2},$$

$$\rho_{\mathbb{H}}(w) = \frac{1}{2\Im w}. \tag{7.14}$$

Thus the map h is an isometry from the upper half plane with its hyperbolic metric, $(\mathbb{H}, \rho_{\mathbb{H}})$, to the unit disk with its hyperbolic metric, (Δ, ρ). Note that this is the same metric as we found in Section 2.3.

Since the map h is an isometry it maps geodesics to geodesics. Since it is a Möbius transformation, it maps colinear points to colinear points so that the geodesics in $\mathbb{H}$ are vertical lines and circles orthogonal to the real axis.

Another example of a simply connected domain where we can find the Riemann map is the Koebe domain K defined as the whole plane minus the part of the negative real axis from $-\infty$ to $-\frac{1}{4}$. The Riemann map is the Koebe map $f_K : \Delta \to K$ given by

$$f_K(z) = \frac{z}{(1-z)^2} = \frac{1}{4}\left(\frac{1+z}{1-z}\right)^2 - \frac{1}{4}.$$

The inverse map $g = f_K^{-1}$ has the formula $g(w) = 1 - \frac{2}{\sqrt{4w+1}+1}$, and by formulas (7.13) and (2.2)

$$\rho_K(w) = \rho(g(w))|g'(w)|,$$

$$\rho_K(w) = \frac{1}{1-|g(w)|^2} \frac{4}{|\sqrt{4w+1}+1|^2|\sqrt{4w+1}|},$$

$$\rho_K(w) = \frac{4}{(|\sqrt{4w+1}+1|^2 - |\sqrt{4w+1}-1|^2)|\sqrt{4w+1}|},$$

$$\rho_K(w) = \frac{1}{\Re\sqrt{4w+1}|\sqrt{4w+1}|}. \tag{7.15}$$

A third example where we can find the Riemann map is the the strip $L = \{w = x + iy \,|\, 0 < y < \lambda \}$. The map $f(z) = \frac{\lambda \log z}{\pi}$ takes the upper half plane $\mathbb{H}$ to L. Therefore, by Corollary 7.3.1,

$$\rho_L(f(z))|f'(z)| = \rho_{\mathbb{H}}(z),$$

and, by formula (7.14),

$$\rho_L(w)\frac{\lambda}{\pi}\frac{1}{|z|} = \frac{1}{2\Im(z)}$$

where $w = f(z)$. Inverting f we have $z = e^{\frac{\pi w}{\lambda}}$ so that

$$\rho_L(w) = \frac{\pi}{2\lambda \sin(\frac{\pi}{\lambda}\Im w)}. \tag{7.16}$$

Example 2 X is a punctured disk.

A punctured disk is a conformal image of the domain Δ^*, the unit disk punctured at zero. Note that the map $f(z) = e^{iz}$ is a universal covering map from the upper half plane $\mathbb{H}$ onto the domain Δ^*. Therefore, Theorem 7.1.1 and formula (7.14) imply

$$\rho_{\Delta^*}(f(z))|f'(z)| = \rho_H(z),$$

$$\rho_{\Delta^*}(e^{iz})|e^{iz}| = \rho_H(z),$$

$$\rho_{\Delta^*}(w)|w| = \rho_H\left(i\log\frac{1}{w}\right),$$

$$\rho_{\Delta^*}(w)|w| = \frac{1}{2\log\frac{1}{|w|}},$$

$$\rho_{\Delta^*}(w) = \frac{1}{2|w|\log\frac{1}{|w|}}. \qquad (7.17)$$

Note that the points i and $i+\pi$ project down to the points $\frac{1}{e}$ and $-\frac{1}{e}$ under the covering map f. Let $\gamma(t)$ be an arc of a circle orthogonal to the real axis, such that $\gamma(t)$ starts at i and ends at $i+\pi$. Set $\hat{\gamma}(t) = \gamma(t) + \pi$. Both $f(\gamma)$ and $f(\hat{\gamma})$ are geodesics in Δ^*. Moreover, each realizes the distance between their endpoints $\frac{1}{e}$ and $-\frac{1}{e}$. Note that we can find other geodesics joining these points that are longer than these (see Exercises 7.6 and 7.7 below). Thus, in $\Delta^*, \rho_{\Delta^*}$ geodesics joining pairs of points are not unique!

Example 3 X is an annulus.

An annulus is a conformal image of the round annulus $A_a = \{t \mid a < |t| < 1\}$, where $0 \le a < 1$. If $a = 0$, then $A_a = \Delta^*$ and the formula for ρ_{A_a} is given in Example 2. Thus, assume $0 < a < 1$. Note that $g(w) = e^{iw}$ maps the strip L to the annulus $A_{e^{-\lambda}}$.

Therefore, for $a = e^{-\lambda}$, by Theorem 7.1.1 and formula (7.16) we have

$$\rho_{A_a}(g(w))|g'(w)| = \rho_L(w) = \frac{\pi}{2\lambda\sin(\frac{\pi}{\lambda}\Im w)}.$$

If we set $t = g(w)$ and substitute we get

$$\rho_{A_a}(t) = \frac{\pi}{2|t|\lambda\sin(\frac{\pi}{\lambda}\log\frac{1}{|t|})}. \qquad (7.18)$$

As $\lambda \to \infty$, A_a tends to the punctured disk. In equation (7.18), for λ large, $\frac{\lambda}{\pi}\sin(\frac{\pi}{\lambda}\log\frac{1}{|t|}) \approx \log\frac{1}{|t|}$ so that $\rho_{A_a} \to \rho_{\Delta^*}$.

Exercise 7.6 Describe all the hyperbolic geodesics that join points $\frac{1}{3}$ and $-\frac{1}{3}$ in Δ^*, the unit disk punctured at 0. Hint: Use the universal covering map and draw the geodesics both in the cover and in Δ^*. See Figure 7.1.

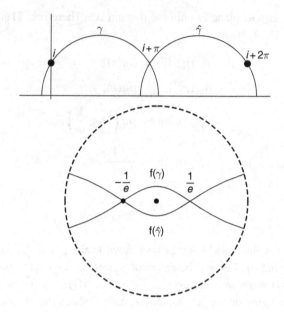

Figure 7.1 Geodesics in the punctured disk

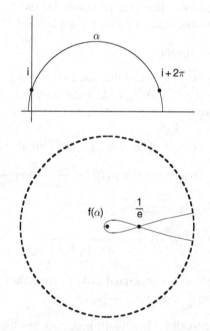

Figure 7.2 Another Geodesic in the punctured disk

Exercise 7.7 Find all the geodesics in the punctured disk Δ^*. Are there any geodesics with self intersections? In particular, if $f(z) = e^{iz}$ and γ is a geodesic in the upper half plane that goes through the points i and $n\pi + i$, $n > 1$, describe the geodesic $f(\gamma)$. See Figure 7.2.

Exercise 7.8 Suppose that X is a simply connected plane domain that contains the point 0 but does not contain the point 1. What are the possible values of $\rho_X(0)$?

Exercise 7.9 Describe the geodesics in the Koebe domain K. See Figure 7.3.

Exercise 7.10 Describe and draw the geodesics in the strip $L = \{z = x + iy \mid 0 < y < 1\}$. Hint: See Figure 7.4.

Exercise 7.11 Let $\Delta^* = \Delta \setminus \{0\}$. Prove that

$$\lim_{|z| \to 1} \frac{\rho_\Delta(z)}{\rho_{\Delta^*}(z)} \to 1.$$

Hint: Use Example 2 and L'Hospital's rule to show

$$\frac{\rho_{\Delta^*}(t_n)}{\rho_\Delta(t_n)} = \frac{\frac{1}{2|t_n|\log\frac{1}{|t_n|}}}{\frac{1}{1-|t_n|^2}} = \frac{1 - |t_n|^2}{2|t_n|\log\frac{1}{|t_n|}} \to 1.$$

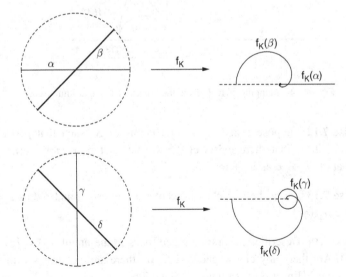

Figure 7.3 Geodesics in the Koebe domain

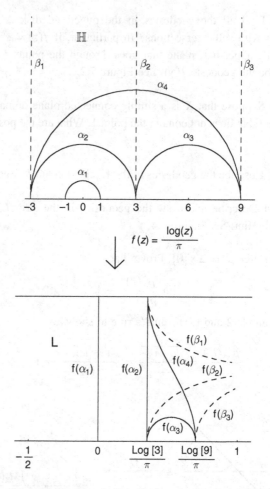

Figure 7.4   Mapping $\mathbb{H}$ to L. Note the branch of the logarithm used.

Exercise 7.12 Suppose z and w are two points in Δ^* such that $\rho(0, z) > n\rho(0, w)$. Show that there exists $c(n) > 0$ such that $\rho_{\Delta^*}(z, w) \geq c(n)$ and such that $c(n) \to \infty$ as $n \to \infty$.

Exercise 7.13 Let γ_c be a circle with center at the origin and radius $c < 1$. Evaluate $\rho_{\Delta^*}(\gamma_c)$.

Exercise 7.14 Describe and draw the geodesics in the annulus $A = \{z \mid 1 < |z| < 3\}$. Are there any closed geodesics? Are there any geodesics with self intersections? Hint: Look at Figures 7.5 and 7.6.

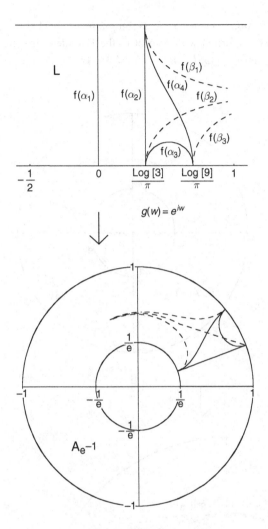

Figure 7.5 Geodesics in the annulus

7.5 Conformal density and curvature

In Section 7.2 we proved Theorem 7.2.1 which says that the hyperbolic density is an infinitesimal form of the hyperbolic distance: that is

$$\rho_X(z, z+t) = |t|\rho_X(z) + o(t).$$

Any positive function λ defining a metric that has this property is called *conformal with respect to the Euclidean metric*. That is:

Definition 7.5 *An arbitrary metric* $\lambda(z, w)$ *defined on* $\Omega \subset \mathbb{R}^2$ *in terms of a density* $\lambda(z)$ *is* conformal with respect to the Euclidean metric, *or simply a* conformal metric, *and the density is called a* conformal density *if*

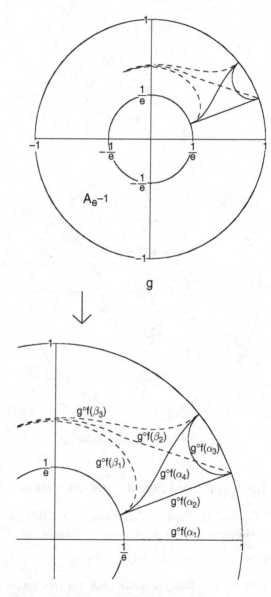

Figure 7.6   Zooming in on the annulus

$$\lim_{t \to 0} \frac{\lambda(z, z+t)}{|t|} = \lambda(z).$$ (7.19)

If λ is twice differentiable, we define

Definition 7.6 *The* Gaussian curvature *of a conformal metric λ_Ω defined on a domain Ω is given by*

$$K(\lambda_\Omega) = -\lambda_\Omega^{-2} \Delta \log \lambda_\Omega.$$

Suppose that f is a conformal map or a regular covering map from Ω onto a domain X and that λ_X is the push-forward of a conformal density λ_Ω; that is, for all $z \in \Omega$,

$$\lambda_X(f(z)) = \frac{\lambda_\Omega(z)}{|f'(z)|}.$$

Moreover, $\Delta(\log |f'(z)|) = 0$, and

$$\Delta_z \log \lambda_\Omega(z) = |f'(z)|^2 \Delta_{f(z)} \log \lambda_X(f(z)).$$

It follows that

$$K(\lambda_\Omega)(z) = K(\lambda_X)(f(z))$$ (7.20)

so that the Gaussian curvature is a conformal invariant of the domain.

We easily compute that the Gaussian curvature with respect to the Euclidean metric is identically zero. For the hyperbolic metric on the disk we can easily compute $K(\rho_\Delta) = K(\rho) = -4$. This proves that the hyperbolic metric has constant negative curvature -4 in the disk. From formula (7.20) we conclude that the hyperbolic metric on any domain has constant negative curvature -4. We remark that what is important here is that the curvature is a negative constant. The value -4 comes from our normalization that $\rho(0) = 1$.

Remark. Had we chosen $\rho(0) = 2$ as is sometimes done, we would have obtained a curvature of -1 and the metric $\rho_\mathbb{H}$ would have taken the form $\rho_\mathbb{H}(z) = \frac{|dz|}{\Im z}$.

7.6 Conformal invariants

7.6.1 Torus invariants

Let T be a torus with a conformal structure. We saw in Section 6.2 that the conformal structure is induced by the conformal structure of the universal cover which can be taken as the plane $\mathbb{C}$. Moreover, T has a metric induced by the Euclidean metric in the plane. Let α and β be two simple closed curves on T that are homotopic neither to a point nor to each other and intersect exactly once at a point $p \in T$; then $[\alpha]$, $[\beta]$ are generators of $\pi_1(T, p)$ and $\pi_1(T, p)$ is isomorphic to the free abelian group on two generators (see Exercise 7.15 below).

Definition 7.7 *The torus T together with a pair of generators $[\alpha]$, $[\beta]$ for $\pi_1(T, p)$ is called a* marked torus *and $[\alpha]$, $[\beta]$ are called* marking generators.

Choose a lift $\tilde{p}$ of p in $\mathbb{C}$. Conjugating $\mathbb{C}$ by a translation we may assume $\tilde{p} = 0$. Now let $\tilde{\alpha}$, $\tilde{\beta}$ be lifts of α and β at $\tilde{p}$. Let $\tilde{p}_\alpha$ be the other endpoint of $\tilde{\alpha}$ and $\tilde{p}_\beta$ the other endpoint of $\tilde{p}_\beta$. Lifting α at $\tilde{p}_\beta$ and β at $\tilde{p}_\alpha$, we obtain a fundamental domain for the covering group. The covering transformations are then represented by $A(z) = z + \tilde{p}_\alpha$ and $B(z) = z + \tilde{p}_\beta$ and $\tilde{p}_\alpha$ and $\tilde{p}_\beta$ are called the periods. That is, we have a representation of the fundamental group with its generators $[\alpha]$, $[\beta]$ as a lattice group in the plane:

$$\sigma : \pi_1(T, p)([\alpha], [\beta]) \to G = \langle A(z), B(z) \rangle.$$

We can "straighten the sides" of the fundamental domain. We replace $\tilde{\alpha}$ by the straight line connecting $\tilde{p}$ and $\tilde{p}_\alpha$, and similarly for the other three sides. In this way, given marking generators for π_1 we signal out a special or "canonical" fundamental domain $P_0(T)$. The universal covering projection sends the straight line sides of the fundamental domain to geodesics on T that are respectively homotopic to α and β.

We can use this canonical domain to prove an easy case of a more general theorem.

Theorem 7.6.1 *Let T be a marked torus with marking generators $[\alpha]$, $[\beta]$. A homeomorphism f from T to a marked torus $f(T)$ with marking generators $[f(\alpha)]$, $[f(\beta)]$ is conformally equivalent to T if and only if the canonical fundamental domains for T and $f(T)$ have the same shape. That is, there are constants $c, d \in \mathbb{C}$ such that, if $S(z) = cz + d$, then $S(P_0(T)) = P_0(f(T))$.*

Proof.   Let π, π_f be the universal covering maps of T and $f(T)$ respectively and define the representation σ_f of $\pi_1(f(T))$ by

$$\sigma_f : \quad f(\alpha) \to A_f(z) = z + \tilde{p}_{f(\alpha)},$$
$$f(\beta) \to B_f(z) = z + \tilde{p}_{f(\beta)}.$$

Given f, we can lift f to a homeomorphism $\tilde{f} : \mathbb{C} \to \mathbb{C}$ by lifting it first to the canonical fundamental domains and then extending using the group. That is, by setting

$$\tilde{f}(z) = \pi_f^{-1} \circ f \circ \pi(z) \text{ where } z \in P_0(T)$$

and π_f^{-1} is chosen so that its image is $P_0(f(T))$. Then, to define it on the rest of $\mathbb{C}$, for any $z' \in \mathbb{C}$ there are $m, n \in \mathbb{Z}$ and some $z \in P_0$ such that $z' = mA_f(z) + nB_f(z)$. Set

$$\tilde{f}(z') = \tilde{f}(mA(z) + nB(z)) = mA_f(\tilde{f}(z)) + nB_f(\tilde{f}(z)).$$

Since the conformal homeomorphisms of the plane are exactly the affine maps the theorem follows. □

This theorem tells us that if f is conformal then

$$\frac{\tilde{p}_{f(\beta)}}{\tilde{p}_{f(\alpha)}} = \frac{\tilde{p}_\beta}{\tilde{p}_\alpha}$$

or, in other words, the ratio of the periods of the lattice is a *conformal invariant* of the marked torus. We give it a special name:

Definition 7.8 *If T is a marked torus represented by a lattice group, the ratio of its periods is called the* modulus of the marked torus *and denoted by* mod(T). *It is uniquely determined if we require its imaginary part to be positive.*

The theorem says that there is a one to one correspondence between points in the upper half plane and conformal equivalence classes of marked tori.

Exercise 7.15 Prove that if α and β are any two simple curves that intersect exactly once on a torus T then $[\alpha]$ and $[\beta]$ generate $\pi_1(T, p)$, where the basepoint p is the intersection point.

7.6.2 Extremal length

In this subsection we turn to the problem of finding invariants for families of curves in arbitrary plane domains. Our discussion is based on [2], Chapter 4, and the reader is referred there for a complete treatment.

Let Ω be a domain in the plane and let Γ be a collection of finite unions of curves in Ω. Given an element $\gamma \in \Gamma$ there is not a unique way to measure its length; we have already seen that we can find both its Euclidean and hyperbolic lengths. In the next chapters we will also define other metrics that can be used to find lengths. All of the metrics we consider are conformal with respect to the Euclidean metric; that is, they can be defined in terms of a density $\sigma(z)|dz|$ where $\sigma(z)$ is a non-negative Borel measurable function. Under a conformal mapping f, with $\Omega' = f(\Omega)$, the density $\sigma(z)$ is transformed into a density $\sigma'(z)$ satisfying

$$\sigma'(f(z))|df(z)| = \sigma(z)|dz|.$$

We can measure the length of $\gamma \in \Gamma$ and the area of Ω with respect to $\sigma(z)$ by

$$L(\gamma, \sigma) = \int_\gamma \sigma(z)|dz|,$$

$$A(\Omega, \sigma) = \int_\Omega \sigma(z)^2 dx dy.$$

These quantities clearly don't change under conformal maps.

Next set

$$L(\Gamma, \sigma) = \inf_{\gamma \in \Gamma} L(\gamma, \sigma).$$

To define an invariant of Ω that doesn't change when σ is multiplied by a constant we define

Definition 7.9 *The* extremal length *of* Γ *in* Ω *is*

$$\lambda_\Omega(\Gamma) = \sup_\sigma \frac{L(\Gamma, \sigma)^2}{A(\Omega, \sigma)}.$$

where the supremum is taken over all conformal densities such that $0 < A(\Omega, \sigma) < \infty$.

Example 4: A *topological rectangle* Q is a domain bounded by a Jordan curve with four marked points $(\alpha, \beta, \gamma, \delta)$, called the corners, occurring in order as the curve is traversed. We will find the extremal length of the family of curves Γ joining one pair of opposite sides of Q; that is, sides with one endpoint in (α, β) and the other in (δ, γ). To find it we first prove

Theorem 7.6.2 *Every topological rectangle* Q *is conformally equivalent to a Euclidean rectangle* R *such that corners are mapped to corners.* R *is unique up to scale and translation.*

Proof. By the Riemann mapping theorem there is a conformal homeomorphism Φ of Q onto the upper half plane $\mathbb{H}$. Since Q is bounded by a Jordan curve, by Theorem 3.4.2, the Riemann mapping Φ extends continuously to the boundary. Post-composing with a Möbius transformation we can choose Φ so that the images of (α, β, δ) are respectively $(0, a, \infty)$ for a given positive a. The image of γ is a point $x \in (a, \infty)$ that depends on Q. There is a holomorphic map g_Q of $\mathbb{H}$ onto a rectangle R_Q with vertices $(0, a, a+ib, ib)$

so that $g_Q(x) = a + ib$ and $g_Q(\infty) = ib$ given by an elliptic integral of the form

$$g_Q(z) = C \int_0^z \frac{dz}{\sqrt{z(z-a)(z-x)}}.$$

The constant C is determined by the condition $g_Q(a) = a$ (see [50], p. 192). The map g_Q is conformal everywhere except at the vertices. It is therefore sufficient to find the extremal length for the rectangle R with vertices $(0, a, a + ib, ib)$ and the family Γ of curves joining the vertical sides.

Using the Euclidean metric, $\sigma = 1$, we see that $L(R, \sigma) = a$ and $A(R, \sigma) = ab$ so that

$$\lambda_R(\Gamma) \geq \frac{a}{b}.$$

Now let σ be arbitrary. Taking γ as a horizontal line we have

$$L(R, \sigma) \leq \int_0^a \sigma dx.$$

Integrating this inequality with respect to dy we have

$$bL(R, \sigma) \leq \int_0^b \int_0^a \sigma \cdot 1 dx dy.$$

By the Cauchy–Schwarz inequality we see that

$$b^2 L(R, \sigma)^2 \leq \int\int_R \sigma^2 dx dy \cdot \int\int_R dx dy = ab A(R, \sigma).$$

It follows that

$$\lambda_R(\Gamma) \leq \frac{a}{b}$$

and we conclude that

$$\lambda_R(\Gamma) = \frac{a}{b}. \quad \square$$

Had we chosen the curves joining the horizontal sides, the same argument would have yielded the reciprocal b/a. Either quantity describes the shape of the rectangle. Following standard convention we call b/a the *modulus* of the rectangle R or, equivalently, the topological rectangle Q. Two topological rectangles (with horizontal and vertical sides marked) are conformally equivalent if and only if they have the same moduli.

Example 5: We let D be a doubly connected plane domain and let Γ be the family of curves joining the boundary components. Again, to find the extremal length we will find a conformal map onto a domain where we can make computations.

Theorem 7.6.3 *Let D be any doubly connected domain. Then there are constants r, R and a round annulus*

$$\mathcal{A} = \{z \mid r < |z| < R\}$$

such that $\mathcal{A}$ is conformally equivalent to D. Any other annulus $\{z \mid r' < |z - a| < R'\}$ with $R'/r' = R/r$ is also conformally equivalent to D.

Proof. The fundamental group $\pi_1(D)$ is generated by the homotopy class of a simple closed curve separating the boundary components.

If D is the whole plane punctured at one point, then, translating that point to the origin, we obtain the doubly infinite annulus with $r = 0$ and $R = \infty$. Otherwise, the domain D is hyperbolic, so that its universal cover π has a representation as the upper half plane and the covering group is a Fuchsian group isomorphic to the fundamental group. As a cyclic group, the Fuchsian group is elementary and either is generated by a parabolic, which we may take as $T(z) = z + 1$, or has a minimal generator $T(z) = \lambda z$, for some $\lambda > 1$ such that every element of the group is of the form $\lambda^n z$ for some integer n.

Now suppose $\lambda = \exp(2\pi^2/\mu)$, $\mu > 0$, $G = \langle T \rangle$, and $\pi : \mathbb{H} \to \mathbb{H}/G = D$. In Examples 1 and 3 we realized the universal covering map of $\mathbb{H}$ onto the annulus $\mathcal{A}$ with $r = e^{-\mu}$ and $R = 1$ by the map $h(z) = \exp \frac{i\mu \log z}{\pi}$. The covering group in this case was G.

We want to use these two covering maps, π and h, to construct a conformal map from D to $\mathcal{A}$. To do this, choose a point z_0 in D and let w_0 be any lift of z_0 by π^{-1}. Then set $f(z_0) = h(w_0)$. For any other point $z \in D$, let γ be a curve joining z_0 to z and let w be the lift of z obtained by lifting γ to a curve $\tilde{\gamma}$ beginning at w_0. Now define $f(z) = h(w)$. To see that the map is well defined, we must show that $f(z)$ is independent of the choice of γ. Suppose γ' is another curve from z_0 to z, not homotopic to γ. Then $\gamma \gamma'^{-1}$ is a closed curve and the endpoints of the lifts $\tilde{\gamma}$ and $\tilde{\gamma}\prime$ are congruent by an element of G. They therefore project to the same point on $\mathcal{A}$.

If T is parabolic, we use the same argument, but a different standard covering map. In Example 2 we realized the universal covering map of $\mathbb{H}$ onto the punctured disk Δ^*, which is the annulus $\mathcal{A}$ with $r = 0$ and $R = 1$, by the function $h(z) = e^{iz}$. We now define a conformal map $f : D \to \Delta^*$ just as we did above.

To see that we may replace the round annulus $\mathcal{A}$ with any round annulus $\mathcal{A}' = \{z \mid r' < |z - a| < R'\}$ whenever $R'/r' = R/r$, we note that the affine map $S(z) = \frac{r'}{r} z$ is a conformal homeomorphism from $\mathcal{A}$ to $\mathcal{A}'$. $\square$

This theorem tells us that any conformal invariant for $\mathcal{A}$ is a conformal invariant for the doubly connected domain D. Furthermore we have

Theorem 7.6.4 *If D is a doubly connected domain equivalent to the annulus $\mathcal{A}$ with inner and outer radii r and R, then the extremal length of the family of curves joining the boundary components is $\frac{1}{2\pi}\log(R/r)$. In particular, two annuli $\mathcal{A} = \{z \mid r < |z - a| < R\}$ and $\mathcal{A}' = \{z \mid r' < |z - a| < R'\}$ are conformally equivalent if and only if $R'/r' = R/r$.*

Proof. Set $\lambda = \log(R/r)$. We may assume that $\mathcal{A} = \{z \mid 1 < |z| < e^\lambda\}$. Let $f(z)$ be the branch of $\log z$ such that $f(1) = 0$. Then $f(\mathcal{A})$ is the rectangle R with vertices $(0, \lambda, \lambda + 2\pi i, 2\pi i)$. The curves joining the inner and outer components of $\mathcal{A}$ that do not intersect the real segment (r, R) map to curves joining the vertical sides of R. Because f is not well defined on this segment, a curve crossing this segment lifts to two segments, one joining the left side to the top (or bottom) of the rectangle, and one joining the bottom (or top) of the rectangle to the right side. Using the Euclidean metric $\rho \equiv 1$ to measure the length of these broken curves the arguments in the proof of Theorem 7.6.2 go through and we conclude that the extremal length of this family is $\frac{\lambda}{2\pi} = \frac{1}{2\pi}\log(R/r)$. $\square$

Above we used the word *modulus* to describe the shape, or conformal invariant, of a torus. We again use the same word to describe the "shape of an annulus". We set $\mathrm{mod}(D) = \frac{2\pi}{\lambda}$ and call it the *modulus* of D. In Chapter 15 we will use annuli with bounded moduli to characterize a useful class of domains.

It follows from the discussion above that any of the quantities λ, $\frac{\lambda}{2\pi}$, and $\exp \lambda$, and even $\log \lambda$, uniquely determines the "shape" of a doubly connected domain. Two doubly connected domains are thus conformally equivalent if and only if they correspond to the same λ.

Note that the modulus of both the punctured disk, $r = 0$, and the punctured plane, $r = 0$, $R = \infty$, is infinity.

7.6.3 General Riemann surfaces

There is a corresponding theory for general Riemann surfaces which is beyond the scope of this book. Suffice it to say that one defines a marking on the surface using generators for the fundamental group. Then one can define a canonical fundamental domain for the group in terms of the marking generators. The general uniformization theorem says:

Theorem 7.6.5 *Let X be a marked Riemann surface. Then X has a representation as the quotient of $\mathbb{H}$ by a Fuchsian group G and the marking determines generators G. A homeomorphism f that preserves the marking is conformal if and only if X and f(X) can be represented by the same Fuchsian group with the same set of generators.*

It is also possible to find geometric quantities that can be read off from the canonical polygon. These are called moduli for the group and form a set of conformal invariants for the marked Riemann surface.

The reader is referred to the relevant literature. See [28], [29] and others.

7.7 The collar lemma

We saw in Section 6.2 that the conformal structure of a torus depends only on the shape of the tile we use to make it and not on its size. We can always assume that its area is 1. This means that if one side of the tile gets very short the other must get long. On the torus, if one generator gets short the other gets long. A similar situation holds for hyperbolic surfaces.

We saw in Theorem 7.2.5 that every simple closed curve that is homotopically non-trivial and not homotopic to a puncture on a hyperbolic surface defines a free homotopy class containing a unique geodesic. For readability, we shall always mean this curve when we talk about simple closed geodesics on a hyperbolic surface.

Let γ be a simple closed geodesic on a hyperbolic surface S.

Definition 7.10 *A collar $C(\gamma)$ on S about γ of (hyperbolic) width w is a doubly connected subdomain of S containing γ defined by*

$$C(\gamma) = \{x \in S \mid \rho_S(x, \gamma) < w/2\}.$$

The collar lemma says that there is always a collar $C(\gamma)$ about the unique geodesic γ in its free homotopy class whose width goes to infinity as the length of γ goes to zero. Precisely,

Lemma 7.7.1 (Collar lemma) *Let γ be a simple closed geodesic on a hyperbolic surface S of length $l_\gamma = l(\gamma)$. Define $w_\gamma = w(\gamma)$ by the relation*

$$\sinh l_\gamma \sinh w_\gamma = 1.$$

Then there is a collar $C(\gamma)$ on S of width w_γ. Moreover, if γ_1 and γ_2 are disjoint closed simple geodesics, then the collars $C(\gamma_1)$ and $C(\gamma_2)$ are disjoint.

Proof. Let S be the given surface and let γ be a simple closed geodesic on S. If S is an annulus (and its fundamental group is elementary) the distance between boundary components is infinite and the whole surface is a collar about γ of infinite width.

Assume therefore that S is not an annulus and that a uniformizing Fuchsian group is non-elementary. To prove the existence of the collar of width w_γ, lift to the universal cover of S and let $A \in G$ be a hyperbolic element whose axis Ax_A projects to γ. If there is not a collar about γ of width w_γ in S, there are a point $x \in \gamma$ and a geodesic ray δ at x that meets γ again before it has gone a distance of w_γ (see Exercise 7.16 below). Let $\tilde{x}$ be a lift of x in Ax_A and let $\tilde{\delta}$ be a lift of δ at $\tilde{x}$. Then $\tilde{\delta}$ meets the axis of another lift $Ax_{A'}$ of γ at distance $d < w_\gamma$.

We may assume that $\tilde{\delta}$ is the real axis and that the axes of A and A' are symmetric about the origin, each at distance $d/2$, as in Figure 7.7. This is clear if $\tilde{\delta}$ meets the axes Ax_A and $Ax_{A'}$ orthogonally since we normalize so that its midpoint is the origin. If it doesn't, let $\tilde{\delta}'$ be the segment of the common orthogonal to the axes of A and A' and let δ' be its projection to S. Since the shortest distance between two geodesics is along the common orthogonal, the length of $\tilde{\delta}'$ and hence the length of δ' are also less than w_γ. Therefore, we may assume that x is the projection of the intersection of $\tilde{\delta}'$

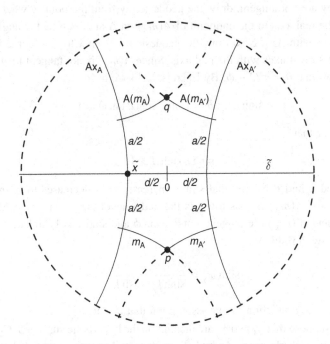

Figure 7.7 There are elliptic points if there is no collar

and Ax_A, and use δ' for the curve δ. We may also assume, using inverses if necessary, that the respective attracting fixed points of A and A' are above the real axis.

Let $a = l_\gamma$ and erect perpendiculars m_A, $m_{A'}$ from Ax_A and $Ax_{A'}$ respectively at distance $a/2$ below the real axis. (See Figure 7.7.) Then $A(m_A)$ and $A'(m_{A'})$ are symmetrically located above the real axis. We claim that m_A and $m_{A'}$ cannot intersect in Δ. Assume that they intersect at $p \in \Delta$. Then, by symmetry, $A(m_A)$ and $A'(m_{A'})$ intersect at a point q. It follows that $A(p) = A'(p) = q$ and $A^{-1}(A')(p) = p$ so that p is a fixed point of an element of G. G, however, cannot have elliptic elements since the covering is regular.

It follows that m_A, $m_{A'}$ (and hence $A(m_A)$ and $A(m_{A'})$) either are tangent or do not meet. If they are tangent, by symmetry, they meet at $-i$ on $\partial\Delta$. We thus have a rectangle with three right angles formed by the real axis $\tilde\delta$, Ax_A, $A(m_A)$ and the geodesic joining i to the origin. The length along the Ax_A side is $a/2$ and the length along the real axis is d. By Exercise 2.15,

$$\sinh a \sinh d = 1,$$

contradicting our assumption that $d < w_\gamma$.

If they are not tangent, draw the geodesic ray from the point o where Ax_A meets the real axis to the endpoint of $A(m_A)$ on Δ and let θ be the angle this ray makes with Ax_A. Next draw the geodesic ray from $\tilde{x}$ to i and let ϕ be the angle this ray makes with the real axis. Since $A(m_A)$ is not tangent to $A(m_{A'})$ it follows that $\theta < \pi/2 - \phi$. By Exercise 2.14 we have

$$\sinh a \tan \theta = 1 \text{ and } \sinh d \tan \phi = 1$$

and hence that

$$\sinh a \sinh d > 1.$$

For fixed a and θ we see that $\sinh d$ increases as θ decreases to 0 and the endpoint of $A(m_A)$ moves towards the endpoint of Ax_A. For $\theta = \pi/2 - \phi$, $A(m_A)$ and $A'(m_{A'})$ are tangent, $\cot\theta = \tan\phi$ and $\sinh d = 1/\sinh a$.

Since by definition

$$\sinh w_\gamma = \frac{1}{\sinh l_\gamma} = \frac{1}{\sinh a}$$

we have a contradiction to our assumption that $d < w_\gamma$.

Now suppose that γ_1 and γ_2 are disjoint on the hyperbolic surface S. Choose lifts to axes of elements A_1 and A_2 such that the distance $d > 0$ between them is as small as possible. Normalize so that the common perpendicular of these axes is the real axis and that its midpoint is the origin. Arguing exactly as above, we can find orthogonals m_{A_1} and m_{A_2}, at points distance $a_1 = l_{\gamma_1}$ on Ax_{A_1} and $a_2 = l_{\gamma_2}$ on Ax_{A_2} below the real axis. Again, an intersection point of these perpendiculars would be a fixed point of $A_1^{-1} A_2$ so they are disjoint. Draw the geodesics joining the terminal points of M_A, $A(M_A)$ and MA^1, $A^1(M_{A^1})$ respectively. They are disjoint and form the collars. It is clear from the picture that

$$d > \frac{w_{\gamma_1}}{2} + \frac{w_{\gamma_2}}{2}. \quad \square \tag{7.21}$$

The collar about a simple curve γ is an annulus C inside the surface S. The following estimate on the modulus of this annulus in terms of the length of γ was given by Maskit, [40]. This modulus is just the extremal length of the family of curves on S homologous to γ which we denote by $\lambda_S(\gamma)$. Since $C \subset S$, and the extremal lengths are defined by taking the infimum over all curves in the family, there are, perhaps, fewer curves in the family on C than the family on S. Therefore

$$\lambda_S(\gamma) \le \lambda_C(\gamma). \tag{7.22}$$

To estimate $\lambda_C(\gamma)$, choose a universal covering map from $\mathbb{H}$ to S such that the closed geodesic γ in the annulus lifts to the imaginary axis and the group element corresponding to γ is $A = \lambda z$, $\lambda > 1$. A lift of the collar is a domain T in $\mathbb{H}$ defined by

$$T = \left\{ z \,\middle|\, 1 < |z| < \lambda, \ \frac{\pi}{2} - \theta < \arg z < \frac{\pi}{2} + \theta \right\}$$

where $0 < \theta < \pi/2$ is determined by

$$\sinh w_\gamma = \tan \theta = \frac{2}{\sqrt{\lambda} - \frac{1}{\sqrt{\lambda}}}.$$

The map $f(z) = \log(-iz)$, $f(i) = 0$ takes $\bar{T}$ onto a rectangle of height 2θ and width $2l_\gamma = \log \lambda$. It follows that $\lambda_C(\gamma) = \frac{l_\gamma}{\theta}$ is the modulus of the rectangle. Then from formula (7.22) we get the estimate

$$\lambda_S(\gamma)\theta \le l_\gamma. \tag{7.23}$$

Exercise 7.16 Show that, if there is not a collar about γ of width $w_\gamma j$ then there are a point x on γ and a geodesic δ that meets γ again before it has gone a distance of w_γ.

Exercise 7.17 Let X be a pair of pants and let γ_1 and γ_2 be simple closed curves respectively homotopic to two of the boundary curves, oriented so that $\gamma_1 \gamma_2$ is simple. Show that any simple closed curve on X is homotopic to either γ_1, γ_2, $\gamma_1 \gamma_2$ or their inverses.

Exercise 7.18 Let Ax_A and Ax_B be disjoint geodesics in the hyperbolic plane, invariant under A and B respectively. Show that the fixed points of the elliptic Möbius transformation whose matrix is $AB - BA$ are endpoints of a geodesic that is perpendicular to both axes Ax_A and Ax_B. Show, moreover, that this is the unique geodesic with this property.

Exercise 7.19 Let Ax_A and Ax_B be disjoint geodesics in the hyperbolic plane, invariant under A and B respectively. Show that, if the fixed points of A are (a, d) and the fixed points of B are (b, c) so that a, b, c, d occur counterclockwise around the boundary of the hyperbolic plane, then the length l of the segment of the common perpendicular between the axes satisfies

$$|cr(a, b, c, d)| = \sinh^2(l) = \frac{\cosh(2l) - 1}{2}.$$

Exercise 7.20 Show that, if three geodesics are disjoint and bound a domain in Δ (no one separates the other two), then their common perpendiculars also bound a domain.

8

The Kobayashi metric

8.1 The classical Kobayashi density

In Chapter 7 we defined the hyperbolic density ρ_X for any hyperbolic domain X using the uniformization theorem and pushing forward the density ρ on the universal cover Δ to X. There is an equivalent definition, introduced by Kobayashi, which, as we will see in the next section, lends itself to generalization.

Let π be a holomorphic covering map from the unit disk Δ to the hyperbolic domain X. If z is any point in X and t is any point in Δ with $\pi(t) = z$, then by Definition 7.2 the hyperbolic metric on X at the point z is defined as

$$\rho_X(z) = \frac{\rho(t)}{|\pi'(t)|}. \tag{8.1}$$

For convenience we introduce the notation $\mathcal{H}ol(X, Y)$ to stand for the family of holomorphic mappings from the source X to the target Y. The contraction property (7.3) implies that any $f \in \mathcal{H}ol(\Delta, X)$, whether or not it is a covering map, satisfies

$$\rho_X(f(t)) \leq \frac{\rho(t)}{|f'(t)|} \tag{8.2}$$

for all points t in Δ. Combining (8.1) and (8.2) we obtain the following formula:

$$\rho_X(z) = \inf \frac{\rho(t)}{|f'(t)|} \tag{8.3}$$

where the infimum is over all $f \in \mathcal{H}ol(\Delta, X)$ and all t in Δ, such that $f(t) = z$.

This leads to the Kobayashi definition of a density for an arbitrary domain X.

Definition 8.1 *The Kobayashi density κ_X for the domain X is defined by the formula*

$$\kappa_X(z) = \inf \frac{\rho(t)}{|f'(t)|} \tag{8.4}$$

where the infimum is over all $f \in \mathcal{H}ol(\Delta, X)$ and all t in Δ, such that $f(t) = z$.

If X is a hyperbolic domain, then, recalling Corollary 7.3.2 in Chapter 7, we see that the Kobayashi density coincides with the hyperbolic density and the universal covering map realizes the infimum. Note that, since we can always pre-compose any covering map π from Δ to X with a conformal self map of the unit disk that maps zero to any given pre-image of the point z in X, we may assume that $t = 0$ in the formulas (8.1) and (8.3). Therefore, we obtain another useful formula for all hyperbolic domains X.

$$\kappa_X(z) = \rho_X(z) = \frac{1}{|\pi'(0)|} = \inf \frac{1}{|f'(0)|} \tag{8.5}$$

where $\pi(0) = z$ and the infimum is over all $f \in \mathcal{H}ol(\Delta, X)$ such that $f(0) = z$.

In Definition 8.1, we did not require that X be a hyperbolic domain. In fact, if X is either the whole plane $\mathbb{C}$, or the plane punctured at a point p, $\mathbb{C} \setminus \{p\}$, the Kobayashi density makes sense.

Theorem 8.1.1 *If X is either the whole plane or the whole plane punctured at a single point, then $\kappa_X(z) = 0$ for any point z in X.*

Proof. Suppose that X is either $\mathbb{C}$ or $\mathbb{C} \setminus \{p\}$. Take any point z in X, let N be a positive integer and set $f(t) = p + (z - p)e^{Nt}$. Then $f \in \mathcal{H}ol(\Delta, X)$ and $f(0) = z$. Formula (8.5) implies

$$\kappa_X(z) \le \frac{1}{|f'(0)|} = \frac{1}{|z - p|N}.$$

Letting N converge to infinity, we see that $\kappa_X(z) = 0$. □

Therefore, whenever the hyperbolic density is not defined, the Kobayashi density is identically equal to zero.

8.2 The Kobayashi density for arbitrary domains

The definition of the Kobayashi density for a hyperbolic domain X depends on the pre-defined hyperbolic density of the fixed domain Δ, the unit disk. Any holomorphic map pushes the density on Δ forward onto X and the Kobayashi density is defined as the smallest possible push-forward. We now study what happens when we change the fixed domain Δ to an arbitrary hyperbolic domain Ω. We fix Ω and call it the *basepoint domain*.

Definition 8.2 *Let X be a domain in the plane. For every z in X let*

$$\kappa_X^\Omega(z) = \inf \frac{\rho_\Omega(t)}{|f'(t)|},$$

where ρ_Ω is the hyperbolic density on Ω and the infimum is over all $f \in \mathcal{H}ol(\Omega, X)$ and all points t in Ω such that $f(t) = z$.

If a holomorphic map f from Ω to X has a critical point at t, as for example in the case where f is a constant map, then by Theorem 7.2.2 the quotient inside the infimum is a positive number divided by zero and, as is usually done in the arithmetic of the extended real line, we take it be positive infinity. Therefore, the range of κ_X^Ω is always contained in the interval $[0, \infty]$. We call κ_X^Ω the *Kobayashi density on X with respect to the given basepoint domain* Ω, or simply the generalized Kobayashi density.

8.2.1 Generalized Kobayashi density: basic properties

The first basic property of the generalized Kobayashi density is the following simple comparison between the ρ_X and κ_X^Ω densities.

Proposition 8.2.1 *For every hyperbolic domain X and every point z in X*

$$\rho_X(z) \leq \kappa_X^\Omega(z). \tag{8.6}$$

Proof. Let f be any holomorphic map from Ω to X with $f(t) = z$. Then f is a weak contraction with respect to the corresponding Poincaré metrics and thus

$$\rho_X(z)|f'(t)| \leq \rho_\Omega(t),$$

$$\rho_X(z) \leq \frac{\rho_\Omega(t)}{|f'(t)|}.$$

Taking the infimum we obtain $\rho_X(z) \leq \kappa_X^\Omega(z)$. $\square$

Corollary 8.2.1 *For all hyperbolic domains X and all z in X, $\kappa_X^\Omega(z) > 0$.*

When the basepoint domain Ω is Δ, we are in the classical case described in the previous section. In that case, by the uniformization theorem, the holomorphic universal covering map onto any hyperbolic domain realizes the infimum.

The second basic property of the generalized Kobayashi density is that, whenever there is a regular holomorphic covering from Ω to the hyperbolic domain X, the generalized Kobayashi density reduces to the classical Kobayashi density.

Proposition 8.2.2 *If there exists a regular holomorphic covering map from Ω onto X then*

$$\kappa_X^\Omega \equiv \rho_X.$$

In particular,

$$\kappa_\Omega^\Omega \equiv \rho_\Omega.$$

Proof. Suppose that π is a regular holomorphic covering map from the basepoint domain Ω to the hyperbolic domain X. Then, by Theorem 7.1.1, π is an infinitesimal isometry; that is, $\rho_\Omega(t) = \rho_X(\pi(t))|\pi'(t)|$ for all t in Ω. Take any point z in X. Since π is surjective, there exists a point t in X such that $\pi(t) = z$. On the one hand, by the definition of the generalized Kobayashi density we have

$$\kappa_X^\Omega(z) \le \frac{\rho_\Omega(t)}{|\pi'(t)|} = \rho_X(z),$$

and on the other, from the first basic property, we have

$$\kappa_X^\Omega(z) \ge \rho_X(z). \quad \square$$

As an immediate corollary we have

Corollary 8.2.2 *If Ω is a simply connected hyperbolic domain then*

$$\kappa_X(z) = \rho_X(z) = \kappa_X^\Omega(z)$$

for all hyperbolic domains X and all z in X.

The third basic property is that, just as for the hyperbolic density, every holomorphic map from one domain X to another domain Y is an infinitesimal weak contraction with respect to the corresponding generalized Kobayashi densities.

Proposition 8.2.3 *If $g \in \mathcal{H}ol(X, Y)$ then*

$$\kappa_Y^\Omega(g(z))|g'(z)| \le \kappa_X^\Omega(z)$$

for all z in X.

Proof. Take $\epsilon > 0$. By the definition of $\kappa_X^\Omega(z)$, there exist a point t in Ω and a holomorphic map f from Ω to X such that $f(t) = z$ and

$$\kappa_X^\Omega(z) \geq \frac{\rho_\Omega(t)}{|f'(t)|} - \epsilon.$$

Therefore, using the map $g \circ f$ in the definition of $\kappa_Y^\Omega(g(z))$, and applying the chain rule, we obtain

$$\kappa_Y^\Omega(g(z))|g'(z)| \leq \frac{\rho_\Omega(t)}{|f'(t)|} \leq \epsilon + \kappa_X^\Omega(z).$$

Letting ϵ tend to zero, we obtain the required inequality. $\square$

Now we integrate the density κ_X^Ω to obtain the generalized Kobayashi distance between any two points in X.

Definition 8.3

$$\kappa_X^\Omega(p, q) = \inf \int_\gamma \kappa_X^\Omega(t)|dt|,$$

where the infimum is over all paths γ in X joining p to q.

Notice that we are again using the same notation for both the density and the distance between points where the distinction again is determined by the number of parameters.

Remark. It may happen that the only holomorphic functions are constants so that $\kappa_X^\Omega \equiv \infty$ and $\kappa_X^\Omega(z, w) = \infty$ for any pair $z, w \in X$. We can interpret this as saying that z and w are in different components of X. Thus, at such points, the topology induced by the generalized Kobayashi density is not equivalent to the hyperbolic (or standard Euclidean) topology of X. Example 1 in the next subsection shows that this can happen.

It may also happen that there are non-constant holomorphic functions from Ω to X, but they all have a critical value at some $z \in X$. At these points $\kappa_X^\Omega(z)$ is infinite. In Markowsky's example, Exercise 8.2 below, there is exactly one non-constant holomorphic function from Ω to X and it has a single critical value at the origin. For this example $\kappa_X^\Omega(w) < \infty$ and $\kappa_X^\Omega(0, w) < \infty$ for any nonzero $w \in X$.

Now we prove that $\kappa_X^\Omega(z, w)$ is actually a complete metric on X with the caveat that $\kappa_X^\Omega(z, w)$ can be infinite.

Proposition 8.2.4 (X, κ_X^Ω) *is a complete metric space.*

Proof.   The symmetry and the triangle inequality follow directly from the definition of the distance between any two points. To show that the distance between different points is nonzero, suppose that z and w are any two distinct points in X. Let ϵ be any positive real number. Pick a rectifiable curve γ in X that is ϵ-close to realizing the infimum so that $\kappa_X^\Omega(z, w) \geq \int_\gamma \kappa_X^\Omega(t)|dt| - \epsilon$. Then, by Proposition 8.2.1,

$$\kappa_X^\Omega(z, w) \geq \int_\gamma \rho_X(t)|dt| - \epsilon \geq \rho_X(z, w) - \epsilon,$$

and, letting ϵ tend to zero,

$$\kappa_X^\Omega(z, w) \geq \rho_X(z, w) > 0. \tag{8.7}$$

Finally, we show completeness. Suppose that a_n is a Cauchy sequence with respect to κ_X^Ω. Inequality (8.7) implies that a_n is a Cauchy sequence with respect to the hyperbolic metric on X, ρ_X. By Theorem 7.1.3, (X, ρ_X) is complete, and so there exists a point a in X such that $\rho_X(a, a_n) \to 0$ as $n \to \infty$. We would like to show that a_n converges to a in the κ_X^Ω metric as well.

For each $\epsilon > 0$, there exists a smallest positive integer $N(\epsilon)$ such that

$$\kappa_X^\Omega(a_n, a_m) \leq \epsilon, \text{ for all } n, m \geq N(\epsilon). \tag{8.8}$$

Hence, taking $\epsilon_n = 1/2^n$, $n = 1, 2, 3, \ldots$, there exists a subsequence $c_n = a_{N(1/2^n)}$ of a_n such that $\kappa_X^\Omega(c_n, c_{n+1}) \leq 1/2^n$. Thus there exists a path γ_n that joins c_n to c_{n+1}, whose length is close to realizing the distance between its endpoints, so that

$$\kappa_X^\Omega(\gamma_n) = \int_{\gamma_n} \kappa_X^\Omega(t)|dt| \leq 1/2^{n-1}.$$

By inequality (8.6), the hyperbolic lengths of the curves γ_n converge to zero. By Corollary 7.2.1, the hyperbolic metric in a neighborhood of a is equivalent to the Euclidean metric, and so the chain γ_N, made by composing all paths γ_n starting with $n = N$, is a path that joins c_N to a and satisfies

$$\kappa_X^\Omega(\gamma_N) = \sum_{n \geq N} \kappa_X^\Omega(\gamma_n) \leq 1/2^{N-2}.$$

Therefore γ_N is a path whose length goes to zero as N goes to infinity. Thus $\kappa_X^\Omega(c_n, a)$ goes to zero and c_n is a κ_X^Ω-convergent subsequence with limit a. This, combined with the inequality (8.8), implies that a_n κ_X^Ω-converges to a as claimed. $\square$

Proposition 8.2.5 *Every holomorphic map is a weak contraction with respect to the generalized Kobayashi metric. That is, if $f \in \mathcal{H}ol(X, Y)$ then $\kappa_Y^\Omega(f(z), f(w)) \leq \kappa_X^\Omega(z, w)$.*

Proof. This is a straightforward consequence of Proposition 8.2.3. □

All of the above properties of the κ_X^Ω density refer to the fixed basepoint domain Ω. The next property tells us what happens when we fix the range domain X and vary the basepoint domain Ω.

Proposition 8.2.6 *(a) If there exists a regular holomorphic covering map π from Ω onto Y, then for all domains X and all points z in X*

$$\kappa_X^\Omega(z) \leq \kappa_X^Y(z).$$

In particular,

(b) if Ω is a simply connected hyperbolic domain, then for all hyperbolic domains X and Y, and all points z in X,

$$\rho_X(z) = \kappa_X(z) = \kappa_X^\Omega(z) \leq \kappa_X^Y(z)$$

and,

(c) if Ω and Y are conformally equivalent domains, then, for all domains X and all points z in X,

$$\kappa_X^\Omega(z) = \kappa_X^Y(z).$$

Proof. Let z be a point in X. Take a small positive constant ϵ. By the definition of $\kappa_X^Y(z)$, there exist $f \in \mathcal{H}ol(Y, X)$ and a point t in Y, such that $f(t) = z$ and

$$\kappa_X^Y(z) \geq \frac{\rho_Y(t)}{|f'(t)|} - \epsilon. \tag{8.9}$$

Since π is a surjective map, there exists a point w in Ω, such that $\pi(w) = t$. Furthermore, since π is a regular holomorphic covering map, by Theorem 7.1.1, π is an infinitesimal isometry from Ω to Y. Therefore,

$$\rho_\Omega(w) = \rho_Y(t)|\pi'(w)|. \tag{8.10}$$

The definition of $\kappa_X^\Omega(z)$ and the chain rule yield

$$\kappa_X^\Omega(z) \leq \frac{\rho_\Omega(w)}{|(f \circ \pi)'(w)|} = \frac{\rho_\Omega(w)}{|f'(t)\pi'(w)|}. \tag{8.11}$$

Combining (8.9), (8.10) and (8.11), we obtain

$$\kappa_X^Y(z) + \epsilon \geq \frac{\rho_Y(t)}{|f'(t)|} = \frac{\rho_\Omega(w)}{|\pi'(w)||f'(t)|} \geq \kappa_X^\Omega(z).$$

Letting ϵ tend to zero, we obtain (a). To show part (c), observe that the conformal map from Ω to Y and its inverse are both regular holomorphic covering maps. Since there exists a universal covering map from the unit disk to any hyperbolic domain, part (b) follows from Corollary 8.2.2, the Riemann mapping theorem and part (a). □

While the hyperbolic density is continuous, for $\kappa_X^\Omega(z)$ we have the weaker statement

Proposition 8.2.7 *The density $\kappa_X^\Omega(z)$ is upper semi-continuous for every domain X.*

Proof. Fix $z_0 \in X$. If $\kappa_X^\Omega(z_0) \neq \infty$, then given $\epsilon > 0$ we can find $f_0 \in \mathcal{Hol}(\Omega, X)$ and a point $w_0 \in \Omega$ with $f_0(w_0) = z_0$ and $f_0'(w_0) \neq 0$ such that

$$\kappa_X^\Omega(z_0) \geq \frac{\rho_\Omega(w_0)}{|f_0'(w_0)|} - \epsilon.$$

Let $z_n \in X$ be a sequence with limit z_0. Since $f_0'(w_0) \neq 0$ we can find a local inverse $g_0 = f_0^{-1}$ and a sequence $w_n = g_0(z_n)$ in Ω such that $w_n \to w_0$. Because f_0' and ρ_Ω are continuous functions, given $\delta > 0$ we can find an N such that, for all $n > N$,

$$\kappa_X^\Omega(z_n) \leq \frac{\rho_\Omega(w_n)}{|f_0'(w_n)|} \leq \frac{\rho_\Omega(w_0)}{|f_0'(w_0)|} + \delta.$$

It follows that

$$\limsup_{n \to \infty} \kappa_X^\Omega(z_n) \leq \frac{\rho_\Omega(w_0)}{|f_0'(w_0)|} + \delta \leq \kappa_X^\Omega(z_0) + \epsilon + \delta.$$

Since ϵ and δ are arbitrarily small, κ_X^Ω is upper semi-continuous as claimed. □

As an immediate corollary we have

Corollary 8.2.3 *The subset of X where $\kappa_X^\Omega = \infty$ is closed.*

8.2.2 Examples

Now we look at what happens in some special cases.

Example 1 Ω is the complex plane punctured at $n \geq 2$ points and X is an arbitrary plane domain whose complement contains at least $n+1$ points.

Suppose that $\Omega = \mathbb{C} \setminus \{p_1, p_2, p_3, \dots, p_n\}$ where $p_1, p_2, p_3, \dots, p_n$ are $n \geq 2$ disjoint points and let $f : \Omega \to X$ be any holomorphic function. Then f is holomorphic in a punctured neighborhood N_i of p_i and has a Laurent series expansion about p_i. If p_i is an essential singularity of f, then, by Theorem 3.1.3, $f(N_i)$ covers every point in $\mathbb{C}$, with the exception of at most one point, infinitely often. Since the complement of X contains at least two points, $f(N_i)$ must cover X and at least one of its boundary points, contradicting the assumption that $f(\Omega) \subset X$.

Therefore, for each i, p_i must be either a pole or a removable singularity and $f|_{N_i}$ can be extended to p_i so that $\tilde{f}(p_i) \in \hat{\mathbb{C}}$ and $\tilde{f}$ is meromorphic. Extending at each point we obtain $\tilde{f}$, whose image is compact and by Theorem 3.1.4 is a single point or the whole sphere with the property $f = \tilde{f}|_\Omega$. Now, if f is not a constant function,

$$X \supset f(\Omega) = \tilde{f}(\Omega) \supset \tilde{f}(\mathbb{C}) \setminus \bigcup_{i=1}^{n} \tilde{f}(p_i).$$

But this cannot be since the complement of X contains at least $n+1$ points. We conclude that the only holomorphic functions $f : \Omega \to X$ are constants and that $\kappa_X^\Omega \equiv \infty$.

Example 2 Ω is an arbitrary hyperbolic domain and X is a simply connected hyperbolic domain.

Since X is a simply connected hyperbolic domain there is a biholomorphic map $h : \Delta \to X$. Applying Proposition 8.2.3 to both h and h^{-1} we deduce $\kappa_X^\Omega(h(z))|h'(z)| = \kappa_\Delta^\Omega(z)$ so we may assume X is Δ.

To evaluate $\kappa_\Delta^\Omega(z)$, recall that the Möbius transformation $A(w) = \frac{w-z}{1-w\bar{z}}$ maps z to 0. Applying Proposition 8.2.3 again,

$$\kappa_\Delta^\Omega(z) = \kappa_\Delta^\Omega(0)|A'(z)| = \frac{\kappa_\Delta^\Omega(0)}{1-|z|^2} = \kappa_\Delta^\Omega(0)\rho(z).$$

We conclude that $\kappa_X^\Omega(z) = K(\Omega)\rho_X(z)$ where $K(\Omega)$ is a constant depending on the domain. By Proposition 8.2.1, $K(\Omega) \geq 1$. We saw in the previous example that $K(\Omega)$ can be infinity. We will see in Example 3 that it can take the value 1.

Suppose now that Ω is a doubly connected domain.

Example 3 Ω is a hyperbolic simply connected domain with a puncture and X is an arbitrary hyperbolic domain.

Since there is a biholomorphic map from Ω union its puncture to Δ, such that the puncture maps to the origin, we may, by Proposition 8.2.6(c), assume $\Omega = \Delta \setminus \{0\} = \Delta^*$. Thus we want to find $\kappa_X^{\Delta^*}(z)$ for any $z \in X$.

Let $f : \Delta^* \to X$ be any holomorphic map. Then, as above, we can find an extension

$$\tilde{f} : \Delta \to X \bigcup \text{ punctures of } X \subset \overline{X}.$$

Fix $z \in X$ and let $\pi : \Delta \to X$ be a universal covering map. Since it is a covering, $\rho_X(\pi(t))|\pi'(t)| = \rho_\Delta(t)$ for all $t \in \Delta$. For the moment fix t such that $\pi(t) = z$.

Choosing $\pi|_{\Delta^*}$ as a candidate f in our definition, we have

$$\kappa_X^{\Delta^*}(z) \le \frac{\rho_{\Delta^*}(t)}{|\pi'(t)|} = \frac{\rho_{\Delta^*}(t)}{\rho_\Delta(t)} \rho_X(z).$$

Now suppose $t_n \in \Delta$ with $|t_n| \to 1$. By pre-composing with a Möbius transformation we can choose $\pi_n : \Delta \to X$ such that $\pi_n(t_n) = z$. From the above we have

$$\kappa_X^{\Delta^*}(z) \le \frac{\rho_{\Delta^*}(t_n)}{\rho_\Delta(t_n)} \rho_X(z).$$

Using Exercise 7.11 we see that

$$\frac{\rho_{\Delta^*}(t_n)}{\rho_\Delta(t_n)} \to 1.$$

We conclude $\kappa_X^{\Delta^*}(z) \le \rho_X(z)$. Combining this with Proposition 8.2.1 we see that $\kappa_X^{\Delta^*}(z) = \rho_X(z)$. In particular, $K(\Delta^*) = 1$.

Exercise 8.1 Show that $K(A) = 1$ where $A = \{z \,|\, 0 < a < |z| < 1\}$. Hint: Estimate $\rho_\Delta(t)/\rho_A(t)$ for $|t|$ close to 1 and use the map $z \mapsto z^n$.

Exercise 8.2 *(G. Markowsky, Personal communication)* Let $K = \{\infty, e^{\frac{\pi i}{6}}, e^{\frac{\pi i}{3}}, e^{\frac{\pi i}{2}}, e^{\pi i}\}$ and set $X = \hat{\mathbb{C}} \setminus K$. Let $f(z) = z^2$ and set $\Omega = \hat{\mathbb{C}} \setminus f^{-1}(K)$; then f maps Ω onto X. Express κ_X^Ω in terms of ρ_Ω and find the set of points in X where κ_X^Ω is infinite.

Exercise 8.3 Evaluate $K(\Omega)$ if Ω is a simply connected domain.

Exercise 8.4 Show that $K(\mathbb{C} \setminus \{0, 1\}) = \infty$.

Exercise 8.5 Evaluate $K(\Delta \setminus \{0, \frac{1}{3}\})$.

9

The Carathéodory pseudo-metric

We now turn to another generalization of the hyperbolic density, the *standard Carathéodory density*.

9.1 The classical Carathéodory density

The standard Kobayashi density κ_Ω was defined in the previous chapter by looking at all holomorphic mappings from the unit disk Δ to X. It is the smallest holomorphic push-forward of the hyperbolic metric in the unit disk. To define the standard Carathéodory density we switch the source and target domains; that is, instead of pushing the hyperbolic metric in Δ forward to the target domain, we pull the hyperbolic metric in Δ back to a given source domain. This means that, instead of considering the holomorphic maps from Δ to X, we consider all holomorphic maps from X to Δ. If f is a holomorphic map from the domain X to Δ, the Schwarz lemma for arbitrary domains implies

$$\rho(f(t))|f'(t)| \leq \rho_X(t) \tag{9.1}$$

for all t in X.

The standard Carathéodory density is based on this inequality; it is the largest pullback of the hyperbolic density in Δ. More precisely,

Definition 9.1 *Let X be a domain in the plane. For every t in X let*

$$c_X(t) = \sup \rho(f(t))|f'(t)|,$$

where ρ is the hyperbolic density on Δ and the supremum is over all $f \in \mathcal{H}ol(X, \Delta)$. The density c_X is called the Carathéodory *density on the domain X.*

163

We remark that, by virtue of equation (8.5), we could equally well have replaced ρ by $\kappa\Delta$ in the definition above. We also note that if X is not hyperbolic then by Liouville's theorem we must have $c_X \equiv 0$. This follows because if X is the whole plane the only candidates for f are constants and if X is the punctured plane $\mathbb{C} \setminus \{p\}$, and if f were a non-constant holomorphic map from X to Δ, the map $f(e^z + p)$ would be a non-constant map from $\mathbb{C}$ to Δ.

If we post-compose the map $f \in \mathcal{H}ol(X, \Delta)$ with a Möbius transformation A fixing Δ such that $f(t) = z$ and $A(z) = 0$ and if we set $g = A \circ f$ we see that

$$\rho(f(t))|f'(t)| = \rho(A(f(t)))|A'(f(t))||f'(t)|$$

so that

$$\rho(f(t))|f'(t)| = \rho(g(t))|g'(t)|.$$

Thus we can assume $f(t) = 0$ in the definition of $c_X(t)$ as we did in the definition of $\kappa_X(z)$ and use the fact that $\rho(f(t)) = 1$. Furthermore,

Proposition 9.1.1 *The Carathéodory density $c_X(t) = \max |f'(t)|$ where the maximum is over all $f \in \mathcal{H}ol(X, \Delta)$ such that $f(t) = 0$.*

Proof. We need to show there is a function $f \in \mathcal{H}ol(X, \Delta)$ such that $f(t) = 0$ for which the supremum is achieved. By the definition of $c_X(t)$, there is a sequence $f_n \in \mathcal{H}ol(X, \Delta)$ such that $f_n(t) = 0$ and $|f_n'(t)| \to c_X(t)$. If X is not hyperbolic, by Liouville's theorem $f_n \equiv 0$, $f \equiv 0$ and $c_X \equiv 0$.

If X is hyperbolic, we can apply Montel's theorem to deduce that f_n is a normal family and some subsequence f_{n_j} converges to a function f that is either constant or open. Moreover, $|f_{n_j}'(t)| \to |f'(t)|$. Now $|f(z)| \leq 1$ because $|f_n(z)| < 1$. If $|f(z)| = 1$ for some $z \in X$, then, by the maximum principle, $f(z)$ is constant with absolute value 1; but this contradicts $f(t) = \lim f_{n_j}(t) = 0$, and we conclude f belongs to $\mathcal{H}ol(X, \Delta)$. $\square$

Inequality (9.1) says that the standard Carathéodory density is always less than or equal to the standard hyperbolic density and thus also by formula (8.5) the standard Kobayashi density; that is, for all z in X,

$$c_X(t) \leq \rho_X(t) = \kappa_X(t). \tag{9.2}$$

Proposition 9.1.2 *For the unit disk Δ all three densities coincide; that is*

$$c_\Delta \equiv \rho_\Delta \equiv \kappa_\Delta.$$

Proof. Applying Definition 9.1 when $X = \Delta$ and plugging in the identity map, $f(t) \equiv t$, we have $c_\Delta(t) \geq \rho(t)$. Applying inequality (9.2), we have $c_\Delta(t) \leq \rho_\Delta(t)$. Since $\kappa_\Delta(t) = \rho(t)$ we get the equality of all three densities. $\square$

An immediate corollary is

Corollary 9.1.1 *If X is a simply connected hyperbolic domain then*

$$c_X \equiv \rho_X \equiv \kappa_X.$$

Proof. If X is a simply connected hyperbolic domain there is a biholomorphic homeomorphism $h : X \to \Delta$. On the one hand, by inequality (9.2), for all $t \in X$,

$$\rho_X(t) = \kappa_X(t) \geq c_X(t).$$

On the other, by Definition 9.1 and the Schwarz lemma,

$$\rho_X(t) = \rho_\Delta(h(t))|h'(t)| \leq c_X(t). \quad \square$$

When X is not simply connected we have

Proposition 9.1.3 *If X is a non-simply-connected hyperbolic domain then the inequality (9.2) is strict; that is, for all $t \in X$,*

$$c_X(t) < \rho_X(t) = \kappa_X(t).$$

Proof. Fix $t \in X$ and let $\pi : \Delta \to X$ be a universal cover of X such that $\pi(0) = t$. Choose $f \in \mathcal{H}ol(X, \Delta)$ such that $f(t) = 0$, $c_X(t) = |f'(t)|$, and set $h = f \circ \pi$. Then $h : \Delta \to \Delta$ with $h(0) = 0$. Note that because X is not simply connected π is not one to one so that neither is h. It follows by the Schwarz lemma that $\rho_\Delta(0)|h'(0)| < \rho_\Delta(0)$. Since π is a universal covering, it is a local isometry so that $\rho_X(t)|\pi'(0)| = \rho_\Delta(0)$.

Combining these relations with Proposition 9.1.1 we obtain

$$c_X(t) = \rho_\Delta(f(t))|f'(t)| = \rho_\Delta(h(0))\frac{|h'(0)|}{|\pi'(0)|} < \frac{\rho_\Delta(0)}{|\pi'(0)|} = \rho_X(t) = \kappa_X(t). \quad \square$$

Example 1 Suppose X is the plane $\mathbb{C}$ from which $n < \infty$ points are removed. It follows as in Example 1, Subsection 8.2.2, that f has an extension to a holomorphic function $\tilde{f} : \mathbb{C} \to \overline{\Delta}$. By Liouville's theorem $\tilde{f}$ and hence f must be constant so again $c_X \equiv 0$.

Exercise 9.1 Find the formula for $c_{\Delta^*}(t)$.

9.2 Generalized Carathéodory pseudo-metric

The standard Carathéodory distance for any plane domain is defined using the family of holomorphic maps from the given domain to the unit disk Δ.

That means that the unit disk together with its hyperbolic density serves as a universal model space, or basepoint domain. To generalize, we change this basepoint domain to an arbitrary hyperbolic plane domain X.

Definition 9.2 *Let Ω be any domain in the plane. For every z in Ω let*

$$c_X^\Omega(z) = \sup \kappa_X(f(z))|f'(z)|,$$

where κ_X is the standard hyperbolic density on X and the supremum is over all $f \in \mathcal{H}ol(\Omega, X)$.

We remark that, by equation (8.5), we could equally well have replaced κ_X by ρ_X in the definition above.

Note that the set in the supremum is never empty because there are always constant holomorphic maps. But, in the case that f is a constant map, or if all maps f have a critical point at z, the product in the supremum is equal to zero. The range of c_X^Ω is always contained in the interval $[0, \infty]$. If X is simply connected, then it is biholomorphically equivalent to the unit disk and the definition of c_X^Ω reduces to the standard definition of the Carathéodory density c_Ω (see Example 2 below). We say that c_X^Ω is *the Carathéodory density on Ω with respect to the given domain X*, or simply the *generalized Carathéodory density*.

9.2.1 Generalized Carathéodory density: basic properties

The first basic property of the generalized Kobayashi density is that it is never less than the Poincaré density. The first basic property of the generalized Carathéodory density is that it is never more than the Poincaré density.

Proposition 9.2.1 *For every hyperbolic domain Ω and every point z in Ω,*

$$c_X^\Omega(z) \le \rho_\Omega(z).$$

Proof. Any $f \in \mathcal{H}ol(\Omega, X)$ is a weak contraction with respect to the corresponding Poincaré metrics:

$$\rho_X(f(z))|f'(z)| \le \rho_\Omega(z).$$

Taking the supremum we obtain $c_X^\Omega(z) \le \rho_\Omega(z)$. □

Corollary 9.2.1 *For all domains Ω and all z in Ω, $c_X^\Omega(z) < \infty$.*

We again explore what happens in the special case when Ω is a covering space for X:

Proposition 9.2.2 *If there exists a regular covering map from Ω onto X, then*

$$c_X^\Omega \equiv \rho_\Omega.$$

In particular,

$$c_\Omega^\Omega \equiv \kappa_\Omega$$

and

$c_X^\Omega \equiv \kappa_\Omega$ *whenever Ω is simply connected and X is hyperbolic.*

Proof. First suppose that π is a covering map from a domain Ω to a domain X. Then, by Theorem 7.1.1, π is an infinitesimal isometry so that $\rho_\Omega(t) = \rho_X(\pi(t))|\pi'(t)|$ for all t in Ω. Since π is a candidate in $\mathcal{H}ol(\Omega, X)$ for the supremum in the definition of the Carathéodory metric, it follows that $\rho_\Omega(t) \leq c_X^\Omega(t)$.

Now Proposition 9.2.1 implies that

$$c_X^\Omega(t) \leq \rho_\Omega(t). \quad \square$$

Corollary 9.2.2 *If Ω is a simply connected hyperbolic domain then*

$$\kappa_\Omega(z) = \rho_\Omega(z) = c_X^\Omega(z),$$

for all hyperbolic domains X and all z in Ω.

The next basic property of the generalized Carathéodory density is that every holomorphic map from one domain Ω to another domain Y is an infinitesimal weak contraction with respect to its generalized Carathéodory density:

Proposition 9.2.3 *If g is a holomorphic map from Ω to Y, then*

$$c_X^Y(g(z))|g'(z)| \leq c_X^\Omega(z),$$

for all z in Ω.

Proof. Take $\epsilon > 0$. By the definition of $c_X^Y(g(z))$, there exists a holomorphic map f from Y to the basepoint domain X, such that

$$c_X^Y(g(z)) - \epsilon \leq \rho_X(f(g(z)))|f'(g(z))|.$$

Therefore, using the map $f \circ g$ in the definition of $c_X^\Omega(z)$, and applying the chain rule, we obtain

$$c_X^\Omega(z) \geq \rho_X(f(g(z)))|f'(g(z))g'(z)| \geq |g'(z)|(-\epsilon + c_X^Y(g(z))).$$

Letting ϵ tend to zero, we obtain the desired conclusion. $\square$

Now we integrate the density c_X^Ω to obtain a generalized Carathéodory pseudo-distance between any two points in Ω.

Definition 9.3

$$c_X^\Omega(p, q) = \inf \int_\gamma c_X^\Omega(t)|dt|,$$

where the infimum is over all paths γ in Ω joining p to q.

It is easy to see that c_X^Ω is a pseudo-metric on Ω; that is, it satisfies all the conditions for a standard metric, except that the distance between two distinct points may equal zero.

Proposition 9.2.4 *Every holomorphic map is a contraction with respect to the generalized Carathéodory pseudo-distance. That is, if f is a holomorphic map from Ω to Y, then $c_X^Y(f(z), f(w)) \le c_X^\Omega(z, w)$.*

Proof. This is a straightforward consequence of Proposition 9.2.3. $\square$

This completes the set of basic properties for the c_X^Ω density. All of these properties refer to the fixed basepoint target domain X. In the next proposition we let the target domain X vary.

Proposition 9.2.5 *(a) If there exists a regular covering map π from X onto Y, then*

$$c_X^\Omega(z) \le c_Y^\Omega(z)$$

for all domains Ω and all points z in Ω. In particular,
(b) if X and Y are conformally equivalent domains, then

$$c_X^\Omega(z) = c_Y^\Omega(z)$$

for all domains Ω and all points z in Ω. In addition,
(c) if X is a simply connected domain not equal to the whole plane, then

$$c_\Omega(z) = c_X^\Omega(z) \le c_Y^\Omega(z)$$

for all hyperbolic domains Ω and Y, and all points z in Ω.

Proof. Let z be a point in Ω. Take a small positive constant ϵ. By the definition of $c_X^\Omega(z)$, there exists a holomorphic map f from Ω to X such that

$$c_X^\Omega(z) - \epsilon \le \rho_X(f(z))|f'(z)|. \tag{9.3}$$

Furthermore, π is a holomorphic covering map, so, by Theorem 7.1.1, π is an infinitesimal isometry from X onto Y. Therefore,

$$\rho_X(f(z)) = \rho_Y(\pi(f(z)))|\pi'(f(z))|. \qquad (9.4)$$

The definition of $c_Y^{\Omega}(z)$ yields

$$c_Y^{\Omega}(z) \geq \rho_Y(\pi(f(z)))|(\pi \circ f)'(z)| = \rho_Y(\pi(f(z)))|f'(z)\pi'(f(z))|. \qquad (9.5)$$

Combining (9.3), (9.4) and (9.5), we obtain

$$c_X^{\Omega}(z) - \epsilon \leq \rho_X(f(z))|f'(z)| = \rho_Y(\pi(f(z)))|\pi'(f(z))||f'(z)| \leq c_Y^{\Omega}(z).$$

Letting ϵ tend to zero, we obtain (a). To prove part (b), observe that the conformal map from X onto Y and its inverse are both holomorphic covering maps. Next, any simply connected domain not equal to the whole plane must be hyperbolic, so there exists a universal covering map from the unit disk to X. Then applying the Riemann mapping theorem and parts (a) and (b) we obtain part (c). □

Combining Proposition 9.2.5 with Propositions 8.2.1 and 9.2.1 we obtain

Corollary 9.2.3 *For all hyperbolic domains Ω and Y, and every $z \in \Omega$,*

$$c_{\Omega}(z) \leq c_X^{\Omega}(z) \leq \rho_{\Omega}(z) \leq \kappa_{\Omega}^X(z).$$

Observe that this corollary is a stronger version of inequality (9.2). We again have semi-continuity rather than continuity.

Proposition 9.2.6 *The density c_X^{Ω} is a lower semi-continuous function on Ω.*

Proof. Fix $z \in \Omega$ and take a sequence $z_n \to z$. By definition, given $\epsilon > 0$ there is a function $f \in \mathcal{H}ol(\Omega, X)$ such that

$$c_X^{\Omega}(z) - \epsilon \leq \rho_X(f(z))|f'(z)|.$$

Since both ρ_X and f' are continuous functions we have

$$\liminf_{n \to \infty} c_X^{\Omega}(z_n) \geq \liminf_{n \to \infty} \rho_X(f(z_n))|f'(z_n)|$$
$$= \rho_X(f(z))|f'(z)| \geq c_X^{\Omega}(z) - \epsilon.$$

Since ϵ is arbitrarily small $c_X^{\Omega}(z)$ is lower semi-continuous as claimed. □

An immediate corollary is

Corollary 9.2.4 *The set on which $c_X^{\Omega}(z) = 0$ is closed.*

Exercise 9.2 Let $A = \{z|0 < a < |z| < 1\}$. Evaluate $c_A(a^{\frac{1}{3}})/c_A(a^{\frac{2}{3}})$.
Hint: Apply Proposition 9.2.3 to the map $g(z) = a/z$.

9.2.2 Examples

Now we look at what happens with the generalized Carathéodory density in the special cases from the previous chapter.

Example 2 Ω is the plane punctured at $n > 2$ points and X is an arbitrary domain with at least $n + 1$ boundary points.

Suppose that $\Omega = \mathbb{C} \setminus \{p_1, p_2, p_3, \ldots, p_n\}$ where $p_1, p_2, p_3, \ldots, p_n$ are $n > 2$ disjoint points in the plane. Any $f \in \mathcal{H}ol(\Omega, X)$ omits more than two points in the plane so cannot have an essential singularity at any of the p_i. As in Example 1 of subsection 8.2.2 the only holomorphic functions from Ω to X are constants and so by Definition 9.2 $c_X^{\Omega} \equiv 0$.

Example 3 Ω is arbitrary and X is simply connected.

For this example, by Proposition 9.2.5 part (c), $c_X^{\Omega} = c_{\Omega}$.

Suppose now that Ω is a doubly connected domain.

Example 4 Ω is a hyperbolic simply connected domain with one puncture and X is arbitrary.

By Proposition 9.2.3 we may assume $\Omega = \Delta^*$. Suppose first that X has no punctures. For any $f \in \mathcal{H}ol(\Omega, X)$ let $\tilde{f}$ be the extension of f to Δ. Then $\tilde{f}(0) \in X$. Thus, for any $t \in \Delta^*$, we have $\rho_X(f(t))|f'(t)| \leq \rho_{\Delta}(t)$, and, taking the supremum,

$$c_X^{\Delta^*}(t) \leq \rho_{\Delta}(t).$$

Now let $\pi : \Delta \to X$ be a universal covering of X and denote by π_X the restriction of π to Δ^*. Then, since $\pi_X \in \mathcal{H}ol(\Delta^*, X)$,

$$c_X^{\Delta^*}(t) \geq \rho_X(\pi_X(t))|\pi_X'(t)| = \rho_{\Delta}(t).$$

Therefore, if X has no punctures, for all $t \in \Delta^*$

$$c_X^{\Delta^*}(t) = \rho_{\Delta}(t).$$

Now suppose X has a puncture. By Proposition 9.2.1 we have $c_X^{\Delta^*}(t) \leq \rho_{\Delta^*}(t)$ for all $t \in \Delta^*$. Let $\pi_X : \mathbb{H} \to X$ be a universal covering of X by the upper half plane $\mathbb{H}$ with covering group G. By Theorem 7.2.5, a property of punctures is that there are simple closed curves homotopic to the puncture that are arbitrarily short. Therefore we can find a sequence of curves γ_t so that, if z_t and $A(z_t)$ are the endpoints of a lift $\tilde{\gamma}_t$, then $\rho_{\mathbb{H}}(z_t, A(z_t))$ is arbitrarily small. Now, by Proposition 2.4.1, A cannot be hyperbolic, so it is a parabolic element in G and, conjugating by a Möbius transformation, we may assume that the parabolic element is $A(z) = z + 1$.

A universal covering map $E : \mathbb{H} \to \Delta^*$ is the map $E(w) = e^{2\pi i w}$ and the covering group H is generated by $A(w) = w + 1$. In this representation H is a subgroup of G so that we can factor $\pi_X = h \circ E$ where $h : \Delta^* \to X$ is a holomorphic map.

Fix $t \in \Delta^*$ and let $w \in \mathbb{H}$ be such that $E(w) = t$ and let $z \in X$ be such that $h(t) = z$. Then we have

$$\rho_X(h(t))|h'(t)| = \rho_X(z)|h'(t)| = \frac{\rho_{\mathbb{H}}(w)}{|E'(w)|} = \rho_{\Delta^*}(t)$$

and taking the supremum $c_X^{\Delta^*}(t) \geq \rho_{\Delta^*}(t)$.

Thus we have shown that if X has punctures $c_X^{\Delta^*}(t) = \rho_{\Delta^*}(t)$.

Note that we used $\mathbb{H}$ rather than Δ here because we can factor $\pi_X = h \circ E$ explicitly.

Exercise 9.3 Let $\Omega = \mathbb{C} \setminus \{2, 3, 4\}$ and $X = \mathbb{C} \setminus \{2, 3\}$. Prove that $c_X^{\Omega}(0) < \frac{1}{8}$.

Exercise 9.4 Let $\Omega = \mathbb{C}$ and let X be any hyperbolic domain. Evaluate $c_X^{\Omega}(z)$.

Exercise 9.5 Let $\Omega = \mathbb{C} \setminus \{p\}$ and let X be any hyperbolic domain. Evaluate $c_X^{\Omega}(z)$ for all $z \neq p$.

Exercise 9.6 Find a formula for $c_X^{\Omega}(t)$, where $\Omega = A \setminus \{\sqrt{a}\}$, $A = \{z \mid 0 < a < |z| < 1\}$ and $X = \Delta$.

Exercise 9.7 Find a formula for $c_X^{\Omega}(t)$, where $\Omega = \Delta^*$ and $X = A$.

Exercise 9.8 *(G. Markowsky, personal communication)* Let $K = \{\infty, e^{\frac{\pi i}{6}}, e^{\frac{\pi i}{3}}, e^{\frac{\pi i}{2}}, e^{\pi i}\}$ and set $X = \hat{\mathbb{C}} \setminus K$. Let $f(z) = z^2$ and set $\Omega = \hat{\mathbb{C}} \setminus f^{-1}(K)$; then f maps Ω onto X. Express c_X^{Ω} in terms of ρ_X and find the set of points in X where c_X^{Ω} vanishes.

10

Inclusion mappings and contraction properties

10.1 Estimates of hyperbolic densities

The hyperbolic metric is an important tool in many branches of mathematics but, as we saw in Chapter 7, except in special cases it is impossible to find an explicit formula for either the hyperbolic density or the hyperbolic distance between two points on an arbitrary domain. Instead, we rely on estimates. One way to obtain estimates is to use holomorphic mappings. After all, as we saw in Corollary 7.3.2 of Chapter 7 and in Section 8.1, we can define the hyperbolic metric as the Kobayashi metric which in turn is defined by looking at families of holomorphic mappings. Now let us look more closely at the connections between holomorphic mappings and the hyperbolic metric.

Suppose that f is a holomorphic mapping from a domain Ω to a domain X. The Schwarz lemma for arbitrary domains, Theorem 7.3.1, implies that f is a local contraction with respect to the corresponding hyperbolic densities and a global contraction with respect to the corresponding hyperbolic metrics. Specifically,

$$\rho_X(f(t))|f'(t)| \leq \rho_\Omega(t) \text{ for all } t \in \Omega \text{ and} \tag{10.1}$$

$$\rho_X(f(t), f(s)) \leq \rho_\Omega(t, s) \text{ for all } t, s \in \Omega. \tag{10.2}$$

Inequality (10.1) is the infinitesimal contraction property and inequality (10.2) is the global contraction property.

Let Ω be a hyperbolic domain for which there is no explicit formula for a universal covering map, and hence no formula for the hyperbolic density on Ω. We can try to compare the hyperbolic density on Ω with the hyperbolic densities of other domains in the plane for which we know the formula using

172

the fact that the inequalities (10.1) and (10.2) imply that bigger domains have smaller hyperbolic density and smaller hyperbolic metric. That is,

$$\rho_X(z) \geq \rho_\Omega(z) \text{ for all } z \in X \text{ whenever } X \subset \Omega. \tag{10.3}$$

In particular, since any hyperbolic domain is contained in the complex plane punctured at two points, the latter has the smallest possible hyperbolic density. There is an explicit formula for this density which we will discuss in Chapter 14.

Now suppose p and q are any two distinct points in the complement of the domain Ω. There is an obvious map from Ω to $C \setminus \{p, q\}$, the inclusion map $i(z) = z$. Therefore, for any z in Ω,

$$\rho_\Omega(z) \geq \sup_{p,q} \rho_{pq}(z),$$

where ρ_{pq} is the hyperbolic density of the plane punctured at p and q, and the supremum is over all pairs of distinct points p and q in the complement of Ω. Note that, because there is an explicit formula for $\rho_{pq}(z)$, this inequality always provides an explicit lower bound on ρ_Ω.

We can also use inequality (10.3) to find explicit upper bounds by choosing nice subdomains, that is, those for which we have an explicit formula for the hyperbolic density. One obvious choice is a disk. Let t be a point in Ω and look at all disks with center at t that are contained in Ω. To get the best possible estimate using this approach, take the largest disk D in Ω with center at t. Since D is a subset of Ω, we have $\rho_\Omega(t) \leq \rho_D(t)$. If d is the Euclidean distance from t to the boundary of Ω, the map $f(z) = t + dz$ is a Riemann map from the unit disk onto D, and $\rho_D(t) = \frac{1}{d}$. Thus,

$$\rho_\Omega(t) \leq \rho_D(t) = \frac{1}{d}.$$

10.2 Strong contractions

A different approach to getting an estimate on the hyperbolic density is to look at the family $\mathcal{H}ol(\Omega, X)$ of holomorphic maps f from a given domain Ω to X, where X is a subdomain of Ω. In this case we have a very interesting situation. In general, the contraction inequalities (10.1) and (10.2) compare two different metrics, a hyperbolic metric on the domain and the hyperbolic metric on the range. For many applications, for example, iterating the map f, or looking for fixed points of f, or whatever, it is much better to work with one fixed metric. We can do this if X is a subdomain of Ω because we can either pre-compose or post-compose with the inclusion map $i(z) = z$ from X into Ω. If we apply the contraction inequalities (10.1) and (10.2) to the compositions $i \circ f$ and $f \circ i$, we obtain

$$\rho_\Omega(f(t))|f'(t)| \le \rho_\Omega(t) \text{ for all } t \in \Omega, \tag{10.4}$$

$$\rho_\Omega(f(t), f(s)) \le \rho_\Omega(t, s) \text{ for all } t, s \in \Omega \tag{10.5}$$

and

$$\rho_X(f(t))|f'(t)| \le \rho_X(t) \text{ for all } t \in X, \tag{10.6}$$

$$\rho_X(f(t), f(s)) \le \rho_X(t, s) \text{ for all } t, s \in X \tag{10.7}$$

respectively. These inequalities say that any holomorphic map from a hyperbolic domain Ω to any subdomain X is both an infinitesimal and a global contraction with respect to both the hyperbolic metric on X and the hyperbolic metric on Ω.

From the inequalities (10.4), (10.5), (10.6) and (10.7) we deduce that the *global Ω-contraction constant*

$$gl_\Omega(f) = \sup_{z, w \in \Omega, \ z \ne w} \frac{\rho_\Omega(f(z), f(w))}{\rho_\Omega(z, w)},$$

the *infinitesimal Ω-contraction constant*

$$l_\Omega(f) = \sup_{z \in \Omega} \frac{\rho_\Omega(f(z))|f'(z)|}{\rho_\Omega(z)},$$

the *global X-contraction constant*

$$gl_X(f) = \sup_{z, w \in X, \ z \ne w} \frac{\rho_X(f(z), f(w))}{\rho_X(z, w)}$$

and the *infinitesimal X-contraction constant*

$$l_X(f) = \sup_{z \in X} \frac{\rho_X(f(z))|f'(z)|}{\rho_X(z)}$$

of the map f are all less than or equal to 1. Furthermore,

Theorem 10.2.1 *The global and infinitesimal contraction constants coincide. That is, $l_\Omega(f) = gl_\Omega(f) \le 1$ and $l_X(f) = gl_X(f) \le 1$.*

Proof. We give the proof for the Ω constants. It is the same for the X constants.

Let z and w be any two points in Ω. If γ is a path that realizes the distance $\rho_\Omega(z, w)$, then

$$\rho_\Omega(f(z), f(w)) \leq \rho_\Omega(f(\gamma)) \leq l_\Omega(f)\rho_\Omega(\gamma) = l_\Omega(f)\rho_\Omega(z, w).$$

Thus, $gl_\Omega(f) \leq l_\Omega(f)$.

To prove the converse, let z be an arbitrary point in Ω and let t be a small positive number. By Theorem 7.2.1, as $t \to 0$,

$$\frac{\rho_\Omega(z, z+t)}{|t|} \to \rho_\Omega(z)$$

and

$$\frac{\rho_\Omega(f(z), f(z+t))}{|t|} = \frac{\rho_\Omega(f(z), f(z+t))}{|f(z) - f(z+t)|} \frac{|f(z) - f(z+t)|}{|t|} \to \rho_\Omega(f(z))|f'(z)|.$$

Therefore, as $t \to 0$,

$$gl_\Omega(f) \geq \frac{\rho_\Omega(f(z), f(z+t))}{\rho_\Omega(z, z+t)} \to \frac{\rho_\Omega(f(z))|f'(z)|}{\rho_\Omega(z)}.$$

This implies $l_\Omega(f) \leq gl_\Omega(f)$. $\square$

For a given holomorphic map $f \in \mathcal{H}ol(\Omega, X)$, all the contraction constants are less than or equal to one. We can ask what conditions on Ω and X imply the contraction constant is uniformly strictly less than one. We say that the family $\mathcal{H}ol(\Omega, X)$ is Ω-*strictly uniform* if

$$l_\Omega = \sup_{f \in \mathcal{H}ol(\Omega, X)} l_\Omega(f) < 1$$

and X-*strictly uniform* if

$$l_X = \sup_{f \in \mathcal{H}ol(\Omega, X)} l_X(f) < 1.$$

In the next sections we will see that families satisfying strictly uniform conditions have other nice properties.

Exercise 10.1 Let $X = \Omega = \Delta$ and $f(z) = z^3$. Evaluate $l_\Delta(f)$.

Exercise 10.2 Let $\Omega = \Delta$ and $X = \Delta^*$. Evaluate l_Δ and l_{Δ^*}.

10.3 Lipschitz domains

The estimates in Section 10.1 were obtained by looking at inclusion mappings from two viewpoints; for the first we looked at inclusion mappings from Ω to domains containing it and for the second we looked at inclusion mappings from subdomains of Ω to Ω. Here we take a more detailed look at the inclusion mappings and their contraction properties.

Let X be a subdomain of Ω, and set $i(z) = z$ for all z in X. The inequalities (10.1) and (10.2) immediately imply that the contraction constant

$$gl(X, \Omega) = \sup_{z,w \in X,\ z \neq w} \frac{\rho_\Omega(z, w)}{\rho_X(z, w)}$$

and the infinitesimal contraction constant

$$l(X, \Omega) = \sup_{z \in X} \frac{\rho_\Omega(z)}{\rho_X(z)}$$

of the map i must be less than or equal to 1.

Furthermore, we have

Theorem 10.3.1 $gl(X, \Omega) = l(X, \Omega) \leq 1$ *whenever X is a subdomain of Ω. If X is not equal to Ω, then the inclusion map i is both a strict contraction for every pair of distinct points and an infinitesimally strict contraction at each point.*

Proof. Set $gl = gl(X, \Omega)$ and $l = l(X, \Omega)$. Let z and w be any two points in X. If γ is a path that realizes the distance $\rho_X(z, w)$, then $\rho_\Omega(z, w) \leq \rho_\Omega(\gamma) \leq l\rho_X(\gamma) = l\rho_X(z, w)$. Thus, $gl \leq l$.

To prove the converse, let z be an arbitrary point in X and let t be a small positive number. By Theorem 7.2.1,

$$\frac{\rho_\Omega(z, z+t)}{|t|} \to \rho_\Omega(z) \text{ as } t \to 0,$$

and

$$\frac{\rho_X(z, z+t)}{|t|} \to \rho_X(z) \text{ as } t \to 0.$$

Therefore,

$$gl \geq \frac{\rho_\Omega(z, z+t)}{\rho_X(z, z+t)} \to \frac{\rho_\Omega(z)}{\rho_X(z)} \text{ as } t \to 0.$$

This implies $l \leq gl$.

To show the second part of this theorem, let z and w be any pair of distinct points in X. Let π_X and π_Ω be universal covering maps from Δ onto X and Ω respectively, with $\pi_X(0) = \pi_\Omega(0) = z$. By the proof of Theorem 7.3.1, the map i lifts to a holomorphic map f from Δ to Δ such that $f(0) = 0$ and

$$\pi_\Omega \circ f = i \circ \pi_X. \tag{10.8}$$

If $\rho_\Omega(z) = \rho_X(z)$, then taking derivatives in (10.8) we obtain $|f'(0)| = 1$. The Schwarz lemma then implies that f is a Möbius transformation. This is impossible since f cannot be surjective because its image contains no points p in Δ with $\pi_\Omega(p) \in \Omega \setminus X$.

Similar reasoning applies if we assume that $\rho_\Omega(z, w) = \rho_X(z, w)$ for some two points z and w in X. By Theorem 7.1.2, there exists a point $t \neq 0$ in the unit disk such that $\pi_X(t) = w$ and $\rho(0, t) = \rho_X(z, w)$. Inequality (10.2), together with equation (10.8), implies $\rho_\Omega(z, w) \leq \rho(0, f(t))$. If $\rho_\Omega(z, w) = \rho_X(z, w)$, then we have

$$\rho(0, t) = \rho_X(z, w) = \rho_\Omega(z, w) \leq \rho(0, f(t)).$$

Again, this inequality together with the Schwarz lemma leads to a contradiction. $\square$

An immediate corollary of Theorems 7.2.2 and 10.3.1 and Exercise 7.4 is

Corollary 10.3.1 *The inclusions from relatively compact subdomains are strict contractions and infinitesimally strict contractions. That is, if X is a relatively compact subdomain of Ω, then $l(X, \Omega) < 1$.*

We want to see what other domains have this property.

Definition 10.1 *A subdomain X of a hyperbolic domain Ω is called* Lipschitz, *or, for short, a* Lip *subdomain of Ω, if the inclusion map from X to Ω is an infinitesimally strict contraction.*

Thus,

Corollary 10.3.2 *Every relatively compact subdomain is Lipschitz.*

Not every subdomain, however, is Lipschitz. For example, if Ω is the unit disk Δ, and X is the punctured unit disk Δ^*, then $l(X, \Omega) = 1$ because, by Exercise 7.11,

$$\frac{\rho(z)}{\rho_{\Delta^*}(z)} \to 1 \text{ as } |z| \to 1. \tag{10.9}$$

The next theorem provides a criterion for a subdomain to be a Lipschitz domain that is often easier to check than the definition. The condition in the theorem was introduced by Beardon *et al.* in [6]. It is a hyperbolic form of the classical *Bloch condition* for domains in the plane which is that the supremum of the radii of all (Euclidean) disks completely contained in the domain is finite.

Define the constant $R(X, \Omega)$ as the supremum of the radii measured with respect to ρ_Ω of hyperbolic disks completely contained in X. Call this constant the *Bloch constant* of the subdomain.[1]

Note that these hyperbolic disks may not be topological disks; for example, a disk of large radius about a point close to the origin in the punctured disk overlaps itself. (See Exercise 10.5 below.)

Definition 10.2 *A subdomain X of Ω is a* Bloch subdomain *if $R(X, \Omega) < \infty$; that is, if its Bloch constant is bounded.*

Note that a fundamental domain of a non-compact hyperbolic surface of finite area is Bloch, but a fundamental domain for a non-compact surface of infinite area is not Bloch.

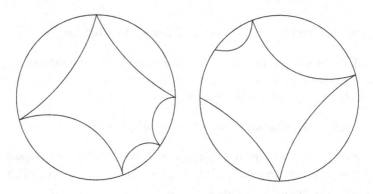

Figure 10.1   The left subdomain is Bloch in Δ and the right is not

Theorem 10.3.2 ([6]) *A subdomain X of Ω is Lipschitz if and only if it is a Bloch subdomain.*

Proof.   Suppose first that $\Omega = \Delta$ and X is a Bloch subdomain of Δ. Then there is a Bloch constant $R(X, \Delta) < \infty$ such that for every $z \in X$ there is a point w in $\Delta \setminus X$ with $\rho(z, w) \leq R(X, \Delta)$. Let $t = A(z)$ where A is a Möbius transformation that sends w to 0. Since $X \subset \Delta \setminus \{w\}$, inequality (10.3) implies

$$\rho_X(z) \geq \rho_{\Delta \setminus \{w\}}(z).$$

Furthermore,

$$\rho(0, t) = \rho(z, w) \leq R(X, \Delta),$$

---

[1] Strictly speaking, we should call this the hyperbolic Bloch constant, but, since we will not use this term in a non-hyperbolic context, we call it the Bloch constant.

which means that t is bounded away from the boundary of the unit circle. In addition we have

$$\frac{\rho(z)}{\rho_X(z)} \leq \frac{\rho(z)}{\rho_{\Delta\setminus\{w\}}(z)} = \frac{\rho(t)}{\rho_{\Delta^*}(t)}.$$

The right side tends to 0 when t tends to 0 so the second half of Theorem 10.3.1 together with Theorem 7.2.2 implies $\frac{\rho(z)}{\rho_X(z)}$ is bounded away from 1. Therefore $l(X, \Delta) < 1$ and X is a Lip subdomain of the unit disk.

Suppose now that X is a Lip subdomain of the unit disk. Let z be an arbitrary point in X. If a hyperbolic disk D with center at z and hyperbolic radius c is contained in X, then inequality (10.3) implies $\rho_X(z) \leq \rho_D(z)$. If D_C is the hyperbolic disk with center at 0, hyperbolic radius c and Euclidean radius C, then $\rho_{D_C}(0) = \frac{1}{C}$. Note that $C \to 1$ as $c \to \infty$. Therefore

$$l(X, \Delta) \geq \frac{\rho(z)}{\rho_X(z)} \geq \frac{\rho(z)}{\rho_D(z)} = \frac{\rho(0)}{\rho_{D_C}(0)} = C.$$

Thus, $C \leq l(X, \Delta) < 1$ and this proves the theorem for $\Omega = \Delta$.

Finally, assume that Ω is an arbitrary hyperbolic domain and let π be a holomorphic universal covering map from the unit disk onto Ω. Let Y be any connected component of $\pi^{-1}(X)$ in Δ. Observe that the restriction of π to the domain Y is a regular covering map onto X. Thus, by Theorem 7.1.1, if z is any point in X and w is any pre-image of z in Y then

$$\frac{\rho_X(z)}{\rho_\Omega(z)} = \frac{\rho_Y(w)}{\rho(w)}.$$

Thus, $l(X, \Omega) = l(Y, \Delta)$ and X is a Lip subdomain of Ω if and only if Y is a Lip subdomain of the unit disk.

Now suppose that z is a point in X and w is any pre-image of z in Y. Let t be a point in $\Omega \setminus X$. Then any $s \in \Delta$ such that $\pi(s) = t$ does not belong to Y. By Theorem 7.1.2, there exists such an s with $\rho(w, s) = \rho_\Omega(z, t)$. It follows that $R(Y, \Delta) \leq R(X, \Omega)$.

Now suppose that p is a point in Y and q is a point in $\Delta \setminus Y$. Let γ be a geodesic in the unit disk that joins p and q. Now $\pi(p)$ belongs to X so there exists a point v on $\pi(\gamma)$ that belongs to the complement of X in Ω. Then,

$$\rho_\Omega(v, \pi(p)) \leq \rho_\Omega(\pi(\gamma)) = \rho(\gamma) = \rho(p, q)$$

and $R(X, \Omega)$ is a lower bound for the Bloch constant $R(Y, \Delta)$. This proves the theorem for any hyperbolic plane domain Ω. $\square$

For more on Lipschitz and Bloch subdomains, see [6].

Exercise 10.3 Let T be an ideal triangle in Δ. Is T a Lip subdomain of Δ ?

Exercise 10.4 Let $S = \{z = x + iy \mid y > 0 \text{ and } -y < x < y\}$ be a sector in the upper half plane $\mathbb{H}$. Is S a Bloch subdomain of $\mathbb{H}$?

Exercise 10.5 Prove that, if z is a point in Δ^* and if r is large enough, the hyperbolic disk about z of radius r overlaps itself.

10.4 Generalized Lipschitz and Bloch domains

Holomorphic functions are contractions with respect to the generalized Kobayashi and Carathéodory densities that we defined in Chapters 8 and 9. In this section we explore the relationships among the various contraction constants.

10.4.1 Kobayashi Lipschitz domains

Let X be any subdomain of a hyperbolic plane domain Ω. By Propositions 8.2.3 and 8.2.5, the inclusion map $i(z) = z$ from X to Ω is both a global and an infinitesimal contraction with respect to the Kobayashi density with basepoint Ω. Thus,

$$\kappa_\Omega^\Omega(t) \le \kappa_X^\Omega(t) \text{ for all } t \in X \tag{10.10}$$

and

$$\kappa_\Omega^\Omega(t, s) \le \kappa_X^\Omega(t, s) \text{ for all } t, s \in X. \tag{10.11}$$

Proposition 8.2.2 implies the global Kobayashi contraction constant

$$gl\kappa(X, \Omega) = \sup_{z,w \in X,\ z \ne w} \frac{\rho_\Omega(z, w)}{\kappa_X^\Omega(z, w)}$$

and the infinitesimal Kobayashi contraction constant

$$l\kappa(X, \Omega) = \sup_{z \in X} \frac{\rho_\Omega(z)}{\kappa_X^\Omega(z)}$$

of the map i are both less than or equal to 1.

In analogy with our definitions for the hyperbolic metric we define

Definition 10.3 *A subdomain X of Ω is a* Kobayashi Lipschitz subdomain *of X, or, for short, a κ-Lip subdomain of Ω, if the inclusion mapping from X to Ω is an infinitesimally strict contraction with respect to the Kobayashi metric with basepoint Ω. That is, a subdomain X of Ω is κ-Lip if and only if $l\kappa(X, \Omega) < 1$.*

Note that Proposition 8.2.1 implies that $l(X, \Omega) \geq l\kappa(X, \Omega)$. Therefore we have the following theorem which shows that the hyperbolic Lipschitz condition is stronger than the Kobayashi Lipschitz condition.

Theorem 10.4.1 *If X is a Lip subdomain of Ω, then X is a κ-Lip subdomain of Ω.*

The following theorem connects this discussion of the generalized Kobayashi metric to our discussion of contraction properties of holomorphic functions.

Theorem 10.4.2 *The Kobayashi contraction constant defined only in terms of the inclusion map and the Ω-contraction constant l_Ω defined in terms of the whole family $\mathcal{H}ol(\Omega, X)$ are equal. That is,*

$$l\kappa(X, \Omega) = l_\Omega.$$

In other words the Kobayashi contraction constant of the inclusion map in $\mathcal{H}ol(X, \Omega)$ is the supremum of the Ω-contraction constants of all holomorphic maps in $\mathcal{H}ol(\Omega, X)$.

Proof of Theorem. Write $l\kappa = l\kappa(X, \Omega)$. For $f \in \mathcal{H}ol(\Omega, X)$, Propositions 8.2.2 and 8.2.3 imply

$$\rho_\Omega(f(z))|f'(z)| \leq l\kappa \, \kappa_X^\Omega(f(z))|f'(z)| \leq l\kappa \, \kappa_\Omega^\Omega(z) = l\kappa \, \rho_\Omega(z).$$

Thus, $l_\Omega(f) \leq l\kappa$. Since this is true for all f, we have $l_\Omega \leq l\kappa$.

To prove the reverse inequality, observe that since $l_\Omega(f) \leq l_\Omega$ for every holomorphic map $f \in \mathcal{H}ol(\Omega, X)$ we deduce

$$\rho_\Omega(f(z)) \leq l_\Omega \frac{\rho_\Omega(z)}{|f'(z)|}. \tag{10.12}$$

Let w be an arbitrary point in X. Taking the infimum of (10.12) over all $f \in \mathcal{H}ol(\Omega, X)$ and all $z \in \Omega$ with $f(z) = w$, we obtain

$$\rho_\Omega(w) \leq l_\Omega \kappa_X^\Omega(w).$$

Thus $l\kappa \leq l_\Omega$. □

As an immediate corollary we see

Corollary 10.4.1 *Uniformly strict contraction with respect to Ω is equivalent to the κ-Lip condition. That is $l\kappa(X, \Omega) < 1$ if and only if $l_\Omega < 1$.*

10.4.2 Kobayashi Bloch domains

We can define the concept of a Bloch domain in the context of the generalized Kobayashi metric and ask if it is equivalent to the κ-Lip condition.

Definition 10.4 *A subdomain X of* Ω *is a* κ-*Bloch subdomain of* Ω *if there exists a constant* $kR(X, \Omega) < \infty$ *such that for every point z in X there is a point w in* $\Omega \setminus X$ *with* $\kappa_\Omega^\Omega(z, w) \leq \kappa R(X, \Omega)$.

By Proposition 8.2.2, however, we immediately have

Theorem 10.4.3 *A domain X is a* κ-*Bloch subdomain of a hyperbolic domain* Ω *if and only if X is a Bloch subdomain of* Ω. *That is, the set of* κ-*Bloch subdomains coincides with the set of Bloch domains.*

Therefore, the κ-Bloch condition for the Kobayashi metric does not provide any more information about subdomains than the Bloch condition for the hyperbolic metric did. In the examples at the end of this chapter we will see that there are κ-Lip domains that are not κ-Bloch so that the κ-Bloch (and equivalently the Bloch and Lip) condition is strictly stronger than the κ-Lip condition. In the special case when Ω is the unit disk, however, we will see in Chapter 13 that all these conditions are equivalent.

10.4.3 Carathéodory Lipschitz domains

We turn now to the generalized Carathéodory metric defined in Chapter 9. Here again we focus on the contraction properties of the inclusion map in $\mathcal{Hol}(\Omega, X)$ but this time with respect to the hyperbolic metric on X. By Propositions 9.2.3 and 9.2.4, the inclusion map $i(z) = z$ from X to Ω is both a global and an infinitesimal contraction with respect to the Carathéodory metric with base X.

Thus,

$$c_X^\Omega(t) \leq c_X^X(t) \text{ for all } t \in X \tag{10.13}$$

and

$$c_X^\Omega(t, s) \leq c_X^X(t, s) \text{ for all } t, s \in X. \tag{10.14}$$

Proposition 9.2.2 says that $c_X^X \equiv \rho_X$, and thus

$$c_X^\Omega(t) \leq \rho_X(t) \text{ for all } t \in X \tag{10.15}$$

and

$$c_X^\Omega(t, s) \leq \rho_X(t, s) \text{ for all } t, s \in X. \tag{10.16}$$

Therefore, the global Carathéodory contraction constant

$$glc(X, \Omega) = \sup_{z, w \in X, \; z \neq w} \frac{c_X^{\Omega}(z, w)}{\rho_X(z, w)}$$

and the infinitesimal Carathéodory contraction constant

$$lc(X, \Omega) = \sup_{z \in X} \frac{c_X^{\Omega}(z)}{\rho_X(z)}$$

of the map i are both less than or equal to 1.

Again, in analogy with our definitions for the hyperbolic metric, we define

Definition 10.5 *A subdomain X of Ω is a* Carathéodory Lipschitz subdomain *of X, or, for short, a* c-Lip subdomain *of Ω, if the inclusion mapping from X to Ω is an infinitesimally strict contraction with respect to the Carathéodory metric with basepoint X. That is, a subdomain X of Ω is c-Lip if and only if $lc(X, \Omega) < 1$.*

Now Proposition 9.2.1 implies that $l(X, \Omega) \geq lc(X, \Omega)$. Therefore we have the following theorem which shows that the hyperbolic Lipschitz condition is stronger than the Carathéodory Lipschitz condition.

Theorem 10.4.4 *If X is a Lip subdomain of Ω, then X is a c-Lip subdomain of Ω.*

The following theorem connects this discussion of the generalized Carathéodory metric to our discussion of contraction properties of holomorphic functions.

Theorem 10.4.5 *The Carathéodory contraction constant defined only in terms of the inclusion map and the X-contraction constant l_X defined in terms of the whole family $\mathcal{H}ol(\Omega, X)$ are equal. That is,*

$$lc(X, \Omega) = l_X.$$

In other words the Carathéodory contraction constant of the inclusion map in $\mathcal{H}ol(X, \Omega)$ is the supremum of the contraction constants of all holomorphic maps in $\mathcal{H}ol(\Omega, X)$.

Proof of Theorem. Write $lc = lc(X, \Omega)$ and let $f \in \mathcal{H}ol(\Omega, X)$. Propositions 9.2.2 and 9.2.3 imply

$$\rho_X(f(z))|f'(z)| = c_X^X(f(z))|f'(z)| \leq c_X^{\Omega}(z) \leq lc\, \rho_X(z).$$

Thus $l_X \leq lc$.

To prove the converse, observe that $l_X(f) \leq l_X \leq 1$ for every $f \in \mathcal{Hol}(\Omega, X)$. This implies

$$\rho_X(f(z))|f'(z)| \leq l_X \rho_X(z). \tag{10.17}$$

Taking the supremum in formula (10.17) over all $f \in \mathcal{Hol}(\Omega, X)$ we obtain

$$c_X^\Omega(z) \leq l_X \rho_X(z).$$

Thus $lc \leq l_X$. $\square$

As an immediate corollary we see

Corollary 10.4.2 *Uniformly strict contraction with respect to X is equivalent to the c-Lip condition. That is $lc(X, \Omega) < 1$ if and only if $l_X < 1$.*

10.4.4 Carathéodory Bloch domains

We can define Bloch domains in the context of the generalized Carathéodory metric and ask how this condition relates to the others we have defined.

Definition 10.6 *A subdomain X of Ω is a c-Bloch subdomain if there exists a constant $cR(X, \Omega) < \infty$ such that for every point z in X there is a point w in $\Omega \setminus X$ with $c_X^\Omega(z, w) \leq cR(X, \Omega)$.*

We will see in the examples below that this condition is not equivalent to the previous two Bloch conditions. Proposition 9.2.1, however, implies

Theorem 10.4.6 *If a domain X is a Bloch subdomain of a hyperbolic domain Ω then X is a c-Bloch subdomain of Ω.*

10.5 Examples

Now we revisit the three examples that we featured earlier as well as one new example.

Example 1 Ω is a plane domain punctured at finitely many points and X is a proper subdomain of Ω.

If X is not equal to Ω, then, as we have seen in Example 1 of both Chapters 8 and 9, $\kappa_X^\Omega \equiv \infty$ and $c_X^\Omega \equiv 0$. Thus

$$l\kappa(X, \Omega) = lc(X, \Omega) = 0 < 1,$$

and so X is both a κ-Lip and a c-Lip subdomain of Ω.

Suppose now that, in addition, X contains a punctured neighborhood of one of the punctures of Ω. That is, there exist a constant $c > 0$ and a puncture p of Ω, such that $D^* = \{z \mid 0 < |z - p| < c\} \subset X$. Then (see Exercise 7.4) the ρ_Ω-distance from a point q in X to the boundary of D^* tends to ∞ as q tends to p. Therefore X is not Bloch in Ω. By Theorem 10.3.2, X is not Lip in Ω.

Since the Carathéodory distance $c_X^\Omega(z, w)$ between any two points in Ω is equal to zero, we have $cR(X, \Omega) = 0$ so that X is a c-Bloch subdomain of Ω. This example shows that the hyperbolic Bloch-Lip condition is strictly stronger than the c-Bloch condition. The implications in Theorems 10.4.1, 10.4.4 and 10.4.6 go in only one direction – the converses do not hold.

Example 2 X is a simply connected subdomain of Ω.

Our discussion of Example 2 in Chapter 8 implies

$$l\kappa(X, \Omega) = \frac{1}{K(\Omega)} l(X, \Omega)$$

where $K(\Omega) = \kappa_\Delta^\Omega(0)$.

Suppose that X is not a Lipschitz subdomain of Ω. Then, X is a κ-Lip subdomain of Ω if and only if $K(\Omega) > 1$.

Example 3 Ω is a simply connected hyperbolic domain with one puncture and X is a subdomain of Ω.

We may assume that $\Omega = \Delta^*$. Example 3 in Chapter 8 implies

$$\kappa_X^{\Delta^*}(z) = \rho_X(z)$$

so that $l(X, \Delta^*) = l\kappa(X, \Omega)$ and X is a Lip subdomain of Δ^* if and only if X is a κ-Lip subdomain of Δ^*.

Suppose first that X has a puncture. Then, by Example 3 in Chapter 9, $c_X^{\Delta^*}(z) = \rho_{\Delta^*}(z)$ and thus $l(X, \Delta^*) = cl(X, \Delta^*)$ and X is a c-Lip subdomain of Δ^* if and only if X is a Lip subdomain of Δ^*.

Suppose now, that X has no punctures. By Example 3 in Chapter 9,

$$c_X^{\Delta^*}(z) = \rho(z)$$

so that $l(X, \Delta) = lc(X, \Delta^*)$ and X is a Lip subdomain of Δ if and only if X is a c-Lip subdomain of Δ^*.

Let us look at a specific example of this kind. Let $z_0 = \frac{1}{2}$ and let z_n, $n = 1, 2, 3, \ldots$, be a sequence of points on the positive real axis such that

$0 < z_n < z_0$ and $\rho_{\Delta^*}(z_0, z_n) = 2^n$. Let I_n be a closed interval on the real axis of ρ_{Δ^*}-length 2 centered at z_n. Let

$$X = \left\{ z \mid 0 < |z| < \frac{1}{2} \right\} \setminus \bigcup_n I_n.$$

Observe that X is relatively compact in Δ, and thus that X is Lip in Δ and c-Lip in Δ^*. Since

$$c_X^{\Delta^*}(z, z_0) = \rho(z, z_0) \le \rho(-z_0, z_0) < \infty,$$

for all $z \in X$, X is c-Bloch in Δ^*. Furthermore, the interval G_n on the real axis located between I_n and I_{n+1} has ρ_{Δ^*}-length $2^n - 2$ so that there are ρ_{Δ^*}-hyperbolic disks in X of arbitrarily large radius and X is not Bloch in Δ^*. By Theorem 10.3.2, X is not Lip in Δ^*. Thus, X is not κ-Lip in Δ^*.

This is an example where the contraction is uniformly strict with respect to X, but not uniformly strict with respect to Δ^*.

Example 4 Now let us look at one further interesting example. For each positive integer n, let D_n be a hyperbolic disk in Δ with center at 0 and ρ-radius n. Let $K_n = \overline{D_n} \setminus D_{n-1}$, and for every point $z \in K_n$ let $D(z, 1)$ be a hyperbolic disk with center at z and radius 1. Since K_n is compact, there is a finite set of points $p_1(n), p_2(n), \ldots, p_{i(n)}(n)$ in K_n such that K_n can be covered by the disks $\{D(p_i(n), 1) \mid 1 \le i \le i(n)\}$.

Let z be any point in the unit disk. Then $z \in K_n$ for some n. Thus there is a point $p_i(n)$ with $\rho(z, p_i(n)) < 1$. Therefore, if

$$\Omega = \Delta \setminus \bigcup_{i,n} p_i(n),$$

then Ω is a plane domain and there is no hyperbolic disk of ρ-radius 1 anywhere in Ω and Ω is a Bloch subdomain of Δ. By Theorem 10.3.2, Ω is also Lip in Δ.

One by one, choose a sequence $\{c_i(n)\}$ so that the disks $D(p_i(n), c_i(n))$ with centers at $p_i(n)$ and radii $c_i(n)$ are pairwise disjoint. Now let $d_i(n)$ be a sequence of closed arcs with center at $p_i(n)$ and ρ-diameter $\frac{c_i(n)}{n}$. Let

$$X = \Omega \setminus \bigcup_{i,n} d_i(n).$$

As a subset of a Bloch subdomain Ω of Δ, X is also Bloch in Δ. For each $n > 2$, let $z_i(n)$ be a point whose ρ-distance to $p_i(n)$ is exactly equal to $c_i(n)/2$. Then

$$\rho(z_i(n), p_j(k)) \ge \frac{c_j(k)}{2}.$$

Therefore, for any k and n, the ρ-distance from $z_i(n)$ to $p_j(k)$ is at least $k/2$ times more than the ρ-diameter of $d_j(k)$. Let w be a point on $\bigcup_{j,k} d_j(k)$ whose ρ_Ω-distance to $z_i(n)$ realizes the minimum. By Exercises 7.4 and 7.12

$$\min_{w \in d_j(k)} \rho_\Omega(z_i(n), w) \geq \min_{w \in d_j(k)} \rho_{\Delta \setminus \{p_j(k)\}}(z_i(n), w) \to \infty \text{ as } n \to \infty.$$

Therefore, X is not Bloch in Ω and, by Theorem 10.3.2, X is not Lip in Ω. Thus,

$$l(X, \Omega) = 1. \tag{10.18}$$

Now let us turn to the Kobayashi and Carathéodory properties. First, we evaluate $\kappa_X^\Omega(z)$. We claim

$$\kappa_X^\Omega(z) = \frac{\rho_X(z)}{l(\Omega, \Delta)}$$

and therefore that X is an example for which the generalized Kobayashi metric is strictly bigger than the hyperbolic metric.

Let z be a point in X, let t be a point in Ω and let $f \in \mathcal{H}ol(\Omega, X)$ satisfy $f(t) = z$. Let $\epsilon > 0$. We may choose t and f such that

$$\kappa_X^\Omega(z) \geq -\epsilon + \frac{\rho_\Omega(t)}{|f'(t)|}. \tag{10.19}$$

Because f is holomorphic and bounded, and Ω has only punctures, f may be extended to a holomorphic map f defined on Δ. By abuse of notation we also call this extension f. Since X has no punctures and f takes a value inside X, the extension f is a map of Δ to X. Inequality (10.1) implies

$$\frac{1}{|f'(t)|} \geq \frac{\rho_X(f(t))}{\rho(t)}. \tag{10.20}$$

Let

$$l(t) = \frac{\rho(t)}{\rho_\Omega(t)}$$

and note that $l(t) \leq l(\Omega, \Delta) < 1$. Plug this and inequality (10.20) into formula (10.19) to obtain

$$\kappa_X^\Omega(z) \geq -\epsilon + \frac{\rho_\Omega(t)}{\rho(t)} \rho_X(z) = -\epsilon + \frac{\rho_X(z)}{l(t)} \geq -\epsilon + \frac{\rho_X(z)}{l(\Omega, \Delta)}.$$

As ϵ tends to 0, f and t vary, and we obtain

$$\kappa_X^\Omega(z) \geq \frac{\rho_X(z)}{l(\Omega, \Delta)}.$$

To prove equality, suppose that t_n is a sequence of points in Ω such that $l(t_n) \to l(\Omega, \Delta)$. Choose a sequence of holomorphic covering maps π_n from Δ onto X such that $\pi_n(t_n) = z$. Then

$$|\pi_n'(t_n)| = \frac{\rho(t_n)}{\rho_X(z)}.$$

Thus,

$$\kappa_X^\Omega(z) \le \frac{\rho_\Omega(t_n)}{|\pi_n'(t)|} = \frac{\rho_\Omega(t_n)}{\rho(t_n)}\rho_X(z) = \frac{\rho_X(z)}{l(t_n)}.$$

Letting n go to infinity proves

$$\kappa_X^\Omega(z) = \frac{\rho_X(z)}{l(\Omega, \Delta)} > \rho_X(z).$$

Therefore,

$$l\kappa(X, \Omega) = \sup_{z \in X} \frac{\rho_\Omega(z)}{\kappa_X^\Omega(z)} = \sup_{z \in X} \frac{\rho_\Omega(z)}{\rho_X(z)} l(\Omega, \Delta) = l(X, \Omega)l(\Omega, \Delta)$$

and equation (10.18) implies

$$l\kappa(X, \Omega) = l(\Omega, \Delta).$$

Since Ω is Lip in the unit disk, we conclude that X is κ-Lip in Ω.

Now we turn to $c_X^\Omega(t)$. We claim the Carathéodory metric on Ω is equal to the hyperbolic metric. That is,

$$c_X^\Omega(t) = \rho(t). \tag{10.21}$$

Inequality (10.20) implies

$$c_X^\Omega(t) \le \rho(t).$$

To prove the converse inequality, take f to be the restriction to Ω of any holomorphic covering map from the unit disk onto X. Then

$$c_X^\Omega(t) \ge \rho_X(f(t))|f'(t)| = \rho(t).$$

Now, since $X \subset \Omega$ and Ω is Lip in Δ,

$$lc(X, \Omega) = \sup_{z \in X} \frac{c_X^\Omega(z)}{\rho_X(z)} = \sup_{z \in X} \frac{\rho_\Delta(z)}{\rho_X(z)} = l(X, \Delta) \le l(\Omega, \Delta) < 1.$$

Therefore X is c-Lip in Ω. Finally, since X is Bloch in Δ, equation (10.21) implies that X is c-Bloch in Ω. Therefore this is another example of a domain that is κ-Lip, c-Lip, c-Bloch but neither Bloch, κ-Bloch nor Lip. In addition, it provides a nice correlation between the contraction constants.

Exercise 10.6 (Open question) Do global and infinitesimal Kobayashi contraction constants coincide for any domain Ω and its subdomain X? That is, does $l\kappa(X, \Omega) = gl\kappa(X, \Omega)$ whenever $X \subset \Omega$?

Exercise 10.7 (Open question) Do global and infinitesimal Carathéodory contraction constants coincide for any domain Ω and its subdomain X? That is, does $lc(X, \Omega) = glc(X, \Omega)$ whenever $X \subset \Omega$?

Exercise 10.8 Let Ω be the unit disk and let

$$X = \left\{ z \mid \left| z - \frac{1}{2} \right| < \frac{1}{2} \right\}.$$

(a) Is X Bloch in Ω?
(b) Is X c-Bloch in Ω?
(c) Is X κ-Lip in Ω?
(d) Is X c-Lip in Ω?

Exercise 10.9 Let Ω be the unit disk and let G be a Fuchsian group with a compact fundamental polygon P. Let V be the set of vertices of P and let

$$X = \Omega \setminus \bigcup_{A \in G} A(V).$$

(a) Is X Bloch in Ω?
(b) Is X c-Bloch in Ω?
(c) Is X κ-Lip in Ω?
(d) Is X c-Lip in Ω?

What if the fundamental domain is not compact but has only parabolic points on the boundary? What if the fundamental domain has finitely many boundary arcs on Ω?

Exercise 10.10 Let Ω be the upper half plane and let

$$X = \{z = x + iy \mid y \geq 1 \text{ and } 0 \leq x \leq 1\}.$$

(a) Is X Bloch in Ω?
(b) Is X c-Bloch in Ω?
(c) Is X κ-Lip in Ω?
(d) Is X c-Lip in Ω?

Exercise 10.11 Let Ω be the punctured unit disk $\Delta^* = \Delta \setminus \{0\}$ and let

$$X = \Omega \setminus \left\{ \frac{1}{2}, \frac{1}{3}, \frac{1}{4}, \cdots \right\}.$$

(a) Is X Bloch in Ω?
(b) Is X c-Bloch in Ω?
(c) Is X κ-Lip in Ω?
(d) Is X c-Lip in Ω?

Exercise 10.12 Let X be any non-relatively-compact subdomain of Δ. Show that there exists a subdomain Y of X such that Y is both Bloch and a non-relatively-compact subdomain of Δ.

11

Applications I: forward random holomorphic iteration

11.1 Random holomorphic iteration

Let Ω be an arbitrary hyperbolic plane domain and let X be any subdomain of Ω. Suppose that we are given a random sequence of holomorphic self maps $f_1, f_2, f_3, \ldots$ from Ω to X. We consider the compositions

$$F_n = f_1 \circ f_2 \circ \cdots \circ f_{n-1} \circ f_n$$

and

$$G_n = f_n \circ f_{n-1} \circ \cdots \circ f_2 \circ f_1.$$

The sequence $\{F_n\}$ is called the *backward iterated function system* arising from the sequence $f_1, f_2, f_3, \ldots$ and the sequence $\{G_n\}$ is called the *forward iterated function system* arising from the sequence $f_1, f_2, f_3, \ldots$ By Montel's theorem, the sequences of functions F_n and G_n both form normal families, and every convergent subsequence converges locally uniformly on compact subsets of Ω to a holomorphic function. The limit functions are called accumulation points. Every accumulation point is either an open map of Ω into itself or a constant map. The constant accumulation points may be located either inside X or on its boundary.

We may look at an iterated function system as a dynamical system acting on Ω. If z is an arbitrary point of Ω, its orbit under the iterated function system, $F_n(z)$ (resp. $G_n(z)$), has $F(z)$ (resp. $G(z)$) as an accumulation point in $\overline{X}$. Hence, if the only limit functions are constants, the orbits of all points tend to periodic cycles. As we will see, we can find conditions so that whether this happens depends only on the geometric properties of the subdomain $X \subset \Omega$ and not on the particular system chosen from $\mathcal{H}ol(\Omega, X)$.

191

In the special case when $\Omega = X = \Delta$ and all maps in the iterated system are the same, forward and backward iteration are the same. The well known Denjoy–Wolff Theorem determines all possible accumulation points.

Theorem 11.1.1 (The Denjoy–Wolff theorem) *Let f be a holomorphic self map of the unit disk Δ that is not a conformal automorphism. Then the iterates $f^{\circ n}$ of f converge locally uniformly in Δ to a constant value t, where $|t| \leq 1$.*

The proof is left as Exercise 12.2.

11.2 Forward iteration

We turn first to forward iteration where $\Omega = \Delta$. This situation is relatively simple as the following theorems show.

Theorem 11.2.1 *Let X be a subdomain of the unit disk Δ. Then all accumulation points of any forward iterated function system of maps in $\mathcal{Hol}(\Delta, X)$ are constant functions if and only if $X \neq \Delta$.*

If X is not relatively compact in Δ, we need the following lemma proved originally in [6].

Lemma 11.2.1 *Suppose X is a proper subdomain of Δ and $a \in X$. Suppose also that $N = N(a, r) \subset X$ is the neighborhood consisting of all points $z \in X$ such that $\rho_X(a, z) < r$. Let f be any holomorphic map from the unit disk Δ into X. If z and w are any two points in Δ such that $f(z)$ and $f(w)$ belong to N then*

$$\rho(f(z), f(w)) \leq C\rho(z, w),$$

where the constant $C = C(N) < 1$ depends only on the neighborhood N.

Proof. Let $N = N(a, 3r)$. By Exercise 2.18 there is a shortest geodesic segment γ in X which joins $f(z)$ and $f(w)$ and stays in N. Now by Exercise 7.4 the closure of N is relatively compact in Δ and, since $X \neq \Delta$, by Theorems 7.2.2 and 10.3.1 there is a constant $C(N) < 1$ such that $\frac{\rho(t)}{\rho_X(t)} \leq C(N)$ for all $t \in N$. Therefore, by Schwarz' lemma we have

$$\rho(f(z), f(w)) \leq \int_\gamma \rho(t)|dt| \leq \int_\gamma C(N)\rho_X(t)|dt|$$

$$= C(N)\rho_X(f(z), f(w)) \leq C(N)\rho(z, w). \quad \square$$

Note that, if X is relatively compact in Δ, then the choice of C is independent of a and N.

We are now ready to prove Theorem 11.2.1.

Proof. Suppose that the sequence G_{n_k} converges locally uniformly on Δ to some holomorphic map G. Since the image of each G_{n_k} is a subset of the closure of X, the same holds for the limit G. Suppose that G is non-constant. Then G is an open map, and so there exists a point z_0 in Δ, such that $G(z_0)$ is in X. If w_0 is any point in Δ with $\rho(z_0, w_0) < 1$, then by the Schwarz lemma each f_i is a weak contraction, and therefore so is G_{n_k}. Thus

$$\rho_X(G_{n_k}(w_0), G_{n_k}(z_0)) \le \rho(z_0, w_0) < 1.$$

It follows that $G_{n_k}(w_0)$ and $G_{n_k}(z_0)$ both belong to $N = N(G(z_0), 2)$ for all sufficiently large k. Lemma 11.2.1 yields

$$\begin{aligned}
\rho(G_{n_k}(z_0), G_{n_k}(w_0)) &\le C(N)\rho(f_{n_k-1} \circ \cdots \circ f_1(z_0), f_{n_k-1} \circ \cdots \circ f_1(w_0)) \\
&\le C(N)\rho(f_{n_k-2} \circ \cdots \circ f_1(z_0), f_{n_k-2} \circ \cdots \circ f_1(w_0)) \\
&\le \cdots \le C(N)\rho(f_{n_{k-1}} \circ \cdots \circ f_1(z_0), f_{n_{k-1}} \circ \cdots \circ f_1(w_0)) \\
&\le C(N)^2 \rho(f_{n_{k-1}-1} \circ \cdots \circ f_1(z_0), f_{n_{k-1}-1} \circ \cdots \circ f_1(w_0)) \\
&\le \cdots
\end{aligned}$$

Therefore $\rho(G_{n_k}(z_0), G_{n_k}(w_0)) \to 0$ as $k \to \infty$ and $G(z_0) = G(w_0)$. This shows that G is constant in a neighborhood of z_0, but, since G is holomorphic, we conclude that G is a constant map as required. $\square$

To complete this section we consider forward iteration for arbitrary hyperbolic domains Ω and subdomains $X \subset \Omega$ such that $X \ne \Omega$. Choose a universal covering map $\pi : \Delta \to \Omega$.

Since $\Omega \ne X$, we can choose a point $z \in \Omega \setminus X$; let $\tilde{z}$ be a fixed pre-image of z under π. Set $Y = \Delta \setminus \{\tilde{z}\}$

For any $f \in \mathcal{H}ol(\Omega, X)$ there is an $\tilde{f} : \Delta \to Y$ such that $\pi \circ \tilde{f} = f \circ \pi$. Since $\pi(\tilde{z})$ is not in X and $f(\Omega) \subset X$, we have $\tilde{f}(\Delta) \subset Y$.

Now let $f_1, f_2, \ldots$ be a random sequence of maps in $\mathcal{H}ol(\Omega, X)$ with

$$G_n = f_n \circ f_{n-1} \circ \cdots \circ f_1$$

and let $\tilde{f}_i$ be a lift of f_i to $\mathcal{H}ol(\Delta, Y)$. The forward iterated sequence

$$\tilde{G}_n = \tilde{f}_n \circ \tilde{f}_{n-1} \circ \cdots \circ \tilde{f}_1$$

has constant limits since Y is a proper subset of Δ. Since

$$\pi \circ \tilde{G}_n = G_n \circ \pi,$$

G_n also has constant limits.

Let K be any closed subset of $\overline{X}$. By Exercise 11.1 below we can find a sequence of points $\{p_n\}$ in X whose accumulation set is precisely K. By taking the iterated function system with each f_n constant and equal to p_n we obtain

Theorem 11.2.2 *Let X be a subdomain of the hyperbolic domain Ω. Then all accumulation points of any forward iterated function system of maps in $\mathcal{H}ol(\Omega, X)$ are constant functions if and only if $X \neq \Omega$. Moreover, the set of accumulation points may be any closed subset of $\overline{X}$.*

Exercise 11.1 Let X be any plane domain and let K be any closed subset of $\overline{X}$. Show that there exists a sequence of points $\{p_n\}$ in X whose accumulation set is precisely the set K. That is, $f \in K$ if and only if there exists a subsequence $\{p_{n_k}\}$ of $\{p_n\}$ that converges to f.
Hint: First take a countable dense subset $C = \{c_1, c_2, c_3, \dots\}$ of K. For every point c_k in C, take a sequence of points c_{k_n} in X such that $|c_{k_n} - c_k| \leq 1/kn$. Finally, take $p_1 = c_{1_1}$, $p_2 = c_{1_2}$, $p_3 = c_{2_1}$, $p_4 = c_{1_3}$, $p_5 = c_{2_2}$, $p_6 = c_{3_1}$ etc.

Exercise 11.2 Let X be a proper subset of the unit disk. If N is any compact subset of X show that there exists a constant $C(N)$ such that $\rho(z, w) \leq C(N)\rho(p, q)$ for any $z, w \in N$, any holomorphic map from Δ to X and any p and q such that $f(p) = z$ and $f(w) = q$.

12

Applications II: backward random iteration

12.1 Compact subdomains

We turn now to backward iterated systems. For readability, from now on we omit the adjective backward. Since we are interested in characterizing those domains for which all limits are constant we give them a special name.

Definition 12.1 *We say that a subdomain X of Ω is* degenerate *if every iterated function system generated by a sequence of maps from Ω to X has only constant accumulation points.*

We begin with a generalization of the Denjoy–Wolff theorem due to Lorentzen and Gill for relatively compact target domains.

Theorem 12.1.1 *([37], [20]) Relatively compact subdomains of Ω are degenerate and any iterated function system has a unique constant limit in X.*

Proof. Suppose that X is relatively compact in Ω. Then, by Corollary 10.3.1, $l(X, \Omega) < 1$. Theorem 10.4.1 implies $l\kappa(X, \Omega) < 1$.

Let $F_n = f_1 \circ f_2 \circ \cdots \circ f_n$ be the iterated function system arising from random functions f_n in $\mathcal{H}ol(\Omega, X)$. If d is the ρ_Ω-diameter of X and if z is any point in Ω, then by Theorem 10.2.1

$$\rho_X(F_n(z), F_{n+m}(z)) \leq \rho_\Omega(f_2 \circ \cdots \circ f_n(z), f_2 \circ \cdots \circ f_{n+m}(z))$$

$$\leq l_\Omega^{n-2} \rho_\Omega(f_n(z), f_n \circ \cdots \circ f_{n+m}(z)) \leq l_\Omega^{n-2} d.$$

Furthermore, by Theorem 10.4.2, $l_\Omega = l\kappa(X, \Omega)$. Therefore $\{F_n(z)\}$ is a Cauchy sequence, and Theorem 7.1.3 implies that $F_n(z)$ converges to a point $F(z)$ in X.

Now, to see that the limit function is constant, suppose w is another point in Ω. Then

$$\rho_\Omega(F_n(z), F_n(w)) \leq l_\Omega^{n-1}\rho_\Omega(f_n(z), f_n(w)) \leq l_\Omega^{n-1}d$$

and, by the triangle inequality, $\rho_\Omega(F(z), F(w)) = 0$. $\square$

12.2 Non-compact subdomains: the $c\kappa$-condition

Beardon, Carne, Minda and Ng showed that there are degenerate subdomains that are not relatively compact. In [6] they prove every Bloch subdomain is degenerate. In this section, we define a more general condition for subdomains and prove that it is a sufficient condition for degeneracy.

For an arbitrary source domain Ω and target subdomain $X \subset \Omega$, we consider the ratio of the following densities. For $z \in X$ define

$$c\kappa_X^\Omega(z) = \frac{c_X^\Omega(z)}{\kappa_X^\Omega(z)}.$$

We have semi-continuity for this ratio.

Proposition 12.2.1 $c\kappa_X^\Omega$ *is lower semi-continuous.*

Proof. Let $z_n \to z_0$. Then by Propositions 8.2.7 and 9.2.6 we have

$$\liminf_{n\to\infty} \frac{c_X^\Omega(z_n)}{\kappa_X^\Omega(z_n)} \geq \frac{\liminf_{n\to\infty} c_X^\Omega(z_n)}{\limsup_{n\to\infty} \kappa_X^\Omega(z_n)} \geq \frac{c_X^\Omega(z_0)}{\kappa_X^\Omega(z_0)} = c\kappa_X^\Omega(z_0). \quad \square$$

Propositions 8.2.1, 9.2.1 and formula (10.3) imply

$$c_X^\Omega(z) \leq \rho_\Omega(z) \leq \rho_X(z) \leq k_X^\Omega(z) \tag{12.1}$$

so that $0 \leq c\kappa_X^\Omega(z) \leq 1$.

Now define

Definition 12.2 *A subdomain X is a $c\kappa$-Lipschitz or $c\kappa$-Lip subdomain of Ω if the inclusion map from X to Ω is a strict contraction with respect to the Kobayashi density on the source and the Carathéodory density on the image. That is, if*

$$\sup_{z\in X} c\kappa_X^\Omega(z) < 1.$$

We can now prove

Theorem 12.2.1 *[31] If X is a cκ-Lipschitz subdomain of Ω, then X is degenerate in Ω.*

Proof. Suppose that X is a $c\kappa$-Lip subdomain of Ω so that

$$\sup_{z \in X} c\kappa_X^\Omega(z) = \sup_{z \in X} \frac{c_X^\Omega(z)}{\kappa_X^\Omega(z)} = C < 1.$$

If X is not degenerate, there are a sequence $f_n \in \mathcal{H}ol(\Omega, X)$ and a subsequence $f_{n_1}, f_{n_2}, \ldots$ such that the subsequence $F_{n_j} = f_1 \circ f_2 \circ \cdots \circ f_{n_j}$ has a non-constant limit function $F(z) : \Omega \to \bar{X}$. Since F is non-constant there is a point $z_0 \in \Omega$ such that $F(z_0) \in X$ and $F'(z_0) \neq 0$.

By the contraction properties of κ_X^Ω and c_X^Ω, Propositions 8.2.3 and 9.2.3, any holomorphic map f from Ω to X satisfies

$$\kappa_X^\Omega(f(z))|f'(z)| \leq \kappa_\Omega^\Omega(z), \tag{12.2}$$

$$c_X^X(f(z))|f'(z)| \leq c_X^\Omega(z). \tag{12.3}$$

These inequalities, together with Propositions 8.2.2 and 9.2.2, imply

$$|f'(z)| \leq \frac{\rho_\Omega(z)}{\kappa_X^\Omega(f(z))} \text{ and } |f'(z)| \leq \frac{c_X^\Omega(z)}{\rho_X(f(z))}. \tag{12.4}$$

Thus

$$|f'(z)|^2 \leq \frac{\rho_\Omega(z)}{\kappa_X^\Omega(f(z))} \frac{c_X^\Omega(z)}{\rho_X(f(z))}. \tag{12.5}$$

We now apply inequality (10.3), the chain rule and formula (12.5) repeatedly, to obtain

$$|F_{n_j}'(z_0)|^2 = |f_1'(f_2 f_3 \ldots f_{n_j}(z_0))|^2 |f_2'(f_3 f_4 \ldots f_{n_j}(z_0))|^2 \ldots |f_{n_j}'(z_0)|^2 \leq$$

$$\frac{\rho_\Omega(f_2 \ldots f_{n_j}(z_0))}{\kappa_X^\Omega(f_1 f_2 \ldots f_{n_j}(z_0))} \frac{c_X^\Omega(f_2 \ldots f_{n_j}(z_0))}{\rho_X(f_1 f_2 \ldots f_{n_j}(z_0))} \frac{\rho_\Omega(f_3 \ldots f_{n_j}(z_0))}{\kappa_X^\Omega(f_2 f_3 \ldots f_{n_j}(z_0))} \frac{c_X^\Omega(f_3 \ldots f_{n_j}(z_0))}{\rho_X(f_2 f_3 \ldots f_{n_j}(z_0))}$$

$$\times \cdots \times \frac{\rho_\Omega(z_0)}{\kappa_X^\Omega(f_{n_j}(z_0))} \frac{c_X^\Omega(z_0)}{\rho_X(f_{n_j}(z_0))} \leq C^{n_j-1} \frac{\rho_\Omega(z_0)}{\rho_X(F_{n_j}(z_0))} \frac{c_X^\Omega(z_0)}{\kappa_X^\Omega(F_{n_j}(z_0))}.$$

Applying inequalites (12.1) we get

$$|F_{n_j}^1(z_0)|^2 \leq C^{n_j-1} \frac{\rho_\Omega(z_0)^2}{\rho_X(F_{n_j}(z_0))^2}.$$

On the one hand $C^{n_j-1} \to 0$, and on the other, by Theorem 7.2.2,

$$\frac{\rho_\Omega(z_0)}{\rho_X(F_{n_j}(z_0))} \to \frac{\rho_\Omega(z_0)}{\rho_X(F(z_0))} < \infty.$$

Therefore $F'(z_0) = 0$, contradicting the assumption that X is not degenerate. $\square$

Finally we have

Proposition 12.2.2 *If X is κ-Lip or c-Lip then it is $c\kappa$-Lip.*

Proof. This follows immediately from formula (12.1). $\square$

Therefore, we also have

Theorem 12.2.2 *([6])* *If X is a Lipschitz subdomain of Ω, then X is degenerate in Ω.*

and

Theorem 12.2.3 *([6])* *If X is a Bloch subdomain of Ω, then X is degenerate in Ω.*

12.3 The overall picture

We have characterized eight different properties for subdomains X of a domain Ω. We want to summarize their relationships. To do so, we divide them into four levels:

(*) Level I: {Lip, Bloch, κ-Bloch}

(*) Level II: {κ-Lip, c-Lip } Level II':   { c-Bloch }

(*) Level III: { $c\kappa$-Lip }

(*) Level IV: { Degenerate }

As a corollary to Theorems 10.3.2 and 10.4.3 we always have

Corollary 12.3.1 *All of the level I properties are equivalent. That is,*

$$X \text{ is Lip} \Leftrightarrow X \text{ is Bloch} \Leftrightarrow X \text{ is } \kappa\text{-Bloch}.$$

As a corollary to Theorems 10.4.6, 10.4.1, 12.2.1 and 10.4.4, Proposition 12.2.2 and Example 1 of Chapter 10, we obtain the following relationships.

Corollary 12.3.2 *Any level I property implies all level II, level III and level IV properties; the level II properties imply the level III and IV properties; and the level III property implies the level IV property; finally the level I properties imply the level II′ property. That is:*

any level I property $\Rightarrow$ *X is* κ*-Lip, c-Lip, c* κ*-Lip and degenerate,*

and these implications are strict;

any level II property $\Rightarrow$ *X is c* κ*-Lip and degenerate;*

the level III property $\Rightarrow$ *X is degenerate;*

and the strict implication

any level I property $\Rightarrow$ *X is c-Bloch.*

It is not known whether the level II′ property implies either level III or level IV.

Example 3 of Chapter 10 shows that, in general, c-Lip does not imply κ-Lip and c-Bloch does not imply κ-Lip. Corollary 12.3.2 implies that if X has any level I, level II or level III property then it is degenerate in Ω. It is an open question whether degeneracy implies the $c\kappa$-Lip condition.

We summarize the relations among the conditions in the following diagram.

$$
\begin{array}{ccccc}
\text{Lipschitz} & \Longleftrightarrow & \text{Bloch} & \Longleftrightarrow & \kappa\text{-Bloch} \\
\nparallel \Downarrow & & \nparallel \Downarrow & & \nparallel \Downarrow \\
\text{c-Lipchitz} & \nRightarrow & \kappa\text{-Lipschitz} & \nLeftarrow & \text{c-Bloch} \\
\Downarrow & & \Downarrow & & \\
& c\kappa\text{-Lipschitz} & & & \\
& \Downarrow & & & \\
& \text{degenerate} & & &
\end{array}
$$

We complete this section by stating

Theorem 12.3.1 *Suppose* Ω *is a simply connected hyperbolic domain and* $X \subset \Omega$*. Let* $\phi : \Omega \to \Delta$ *be a Riemann map and set* $Y = \phi(X)$*. Then X as a subdomain of* Ω *has any one of the eight properties if and only if Y as a subdomain of* Δ *has the same property.*

The proof is left as a set of exercises. In the next chapter we prove a number of theorems in which we assume the source domain is the unit disk. Theorem 12.3.1 implies that they all hold equally well for any simply connected hyperbolic plane domain.

Exercise 12.1 Let f be a holomorphic self map of the unit disk Δ that is not an elliptic Möbius transformation. Suppose that the orbit $f^{\circ n}(0)$ visits a fixed compact set $K \subset \Delta$ infinitely often. Show that the accumulation points of the iterated function system $F_n = f^{\circ n}$ are constants inside Δ.

Exercise 12.2 (The Denjoy–Wolff theorem) Let f be a holomorphic self map of the unit disk Δ that is not an elliptic Möbius transformation. Show that the iterated function system $F_n = f^{\circ n}$ has a constant accumulation point t, where $|t| \le 1$. Hint: You may want to read Chapter 13 first.

Exercise 12.3 Prove Theorem 12.3.1 for the level I properties.

Exercise 12.4 Prove Theorem 12.3.1 for the level II and level II' properties.

Exercise 12.5 Prove Theorem 12.3.1 for the level III and IV properties.

Exercise 12.6 Show that whenever $z \in X \subset \Omega \neq X$ then $c\kappa_X^\Omega(z) < 1$.

13

Applications III: limit functions

In this chapter we look at non-compact subdomains of the unit disk and show that, even if they are degenerate (Bloch), there are iterated function systems with more than one constant limit function.

We then show that, if the subdomain is non-Bloch, there are iterated function systems with non-constant limits and that, in fact, any function may be realized as the limit of an iterated function system.

By the Riemann mapping theorem and Theorem 12.3.1 the results in this chapter hold for subdomains of any simply connected hyperbolic domain.

13.1 Uniqueness of limits

In this section we show that for any integer $n > 1$, and any non-relatively-compact subdomain $X \subset \Delta$, there is an iterated function system that has n arbitrarily chosen distinct accumulation points. We prove

Theorem 13.1.1 *[34] Let X be any subdomain of Δ that is not relatively compact and let $c_0, c_1, \ldots, c_{n-1}$ be n distinct points in X. There is an iterated function system that has at least n distinct accumulation points $G_0, G_1, \ldots, G_{n-1}$ and $G_i(0) = c_i$, $i = 0, \ldots, n-1$. If X is Bloch these accumulation points are constant and there are no other accumulation points.*

13.1.1 The key lemma

In this section we prove a lemma that is the crux of the proof of Theorem 13.1.1.

Lemma 13.1.1 *Let X be any non-relatively-compact subset of Δ, and, for any fixed n, let $a_1, \ldots, a_n$ be any distinct points in $\Delta \setminus \{0\}$. Then there exist a degree-$(n-1)$ rational function $f : \Delta \to \Delta$ and points $x_1, \ldots, x_n \in X$ such that, for all $i = 1, \ldots, n$, $f(x_i) = a_i/x_i$.*

201

Proof. We use the notation

$$A(a, z) = \frac{z - a}{1 - \bar{a}z}$$

and note that $A(a, A(-a, z)) = z$.

Step 1: Since X is not relatively compact we may choose an $x_1 \in X$ such that $|x_1| > |a_1|$. Let $g_1(z)$ be a self map of the unit disk to be determined. Define

$$f(z) = \frac{A(x_1, z)g_1(A(x_1, z)) + \frac{a_1}{x_1}}{1 + \frac{\bar{a}_1}{\bar{x}_1}A(x_1, z)g_1(A(x_1, z))}.$$

It follows that $f(x_1) = a_1/x_1$ as required. Because we want to work inductively we rewrite this definition implicitly as follows:

$$A(x_1, z)g_1(A(x_1, z)) = A\left(\frac{a_1}{x_1}, f(z)\right). \tag{13.1}$$

If $n = 1$ we set $g_1(z) \equiv 0$ and we are done. From now on we assume that $n > 1$ and that we have chosen x_1.

Step 2: Before we proceed, we set up some further notation:
For $1 \leq j \leq k \leq n$, set $a_{jk} = A(x_j, x_k)$. Next, for $k = 2, \ldots, n$ set

$$b_{1k} = A\left(\frac{a_1}{x_1}, \frac{a_k}{x_k}\right). \tag{13.2}$$

For $j = 2, \ldots, n - 1$ and $k = j, j + 1, \ldots, n$ set

$$b_{jk} = A\left(\frac{b_{(j-1)j}}{a_{(j-1)j}}, \frac{b_{(j-1)k}}{a_{(j-1)k}}\right). \tag{13.3}$$

In order that our construction work we need to choose the x_i so that the following inequalities hold:

$$\left|\frac{a_i}{x_i}\right| < 1, \quad i = 1, \ldots, n. \tag{13.4}$$

In Step 1 we chose x_1 so this holds for $i = 1$.
For all j, k such that $j < k$ we also need to have

$$\left|\frac{b_{jk}}{a_{jk}}\right| < 1. \tag{13.5}$$

To see that we can choose the x_i's so that these inequalities are all satisfied. Note first that for fixed j, and all $k > j$, $|x_k| \to 1$ implies $|a_{jk}| \to 1$.
Next, as $|x_j| \to 1$,

$$\limsup |b_{1j}| \leq \left| A\left(\frac{a_1}{x_1}, a_j e^{\theta_j}\right) \right| = B_{1j} < 1,$$

where θ_j is chosen so that $\arg a_j e^{\theta_j} = \arg \frac{a_1}{x_1} + \pi$ and B_{1j} is maximal.

Choose x_1 so that, if the remaining $|x_i|$ are close enough to 1, inequalities (13.4) and (13.5) hold with $j = 1$.

We now find bounds

$$\limsup_{|x_j| \to 1} |b_{2j}| \le \left| A\left(\frac{b_{12}}{a_{12}}, b_{1j}e^{\theta_j}\right) \right| = B_{2j} < 1$$

where again θ_j is chosen to maximize B_{2j}.

We repeat this process, choosing $x_3, \ldots, x_{n-1}, x_n$ in turn so that all the inequalities above hold.

Step 3: Define the functions $g_k(z) : \Delta \to \Delta$, $k = 2, \ldots, n$, recursively by

$$A(x_k, z)g_k(A(x_k, z)) = A\left(\frac{b_{(k-1)k}}{a_{(k-1)k}}, g_{(k-1)}(A(x_{(k-1)}, z))\right). \tag{13.6}$$

Now take $g_n(z)$ to be the function $g_n(z) \equiv 0$. Then work back through equations (13.6) to obtain the functions g_1 and f. Note that g_1 is a rational function of degree $n - 2$ and f is a rational function of degree $n - 1$.

Check that $f(x_i) = \frac{a_i}{x_i}$ for $i = 1, \ldots, n$, so that the points x_i and the function f are as required by the lemma. $\square$

13.1.2 Proof of Theorem 13.1.1

Now we are ready to prove Theorem 13.1.1. The functions of the iterated function system are constructed inductively.

Proof. With no loss of generality we may assume that $0 \in X$. The idea of the proof is to construct functions f_k such that the set $S = \{c_0 = 0, c_1 = F_1(0) = f_1(0), c_2 = F_2(0) = f_1 \circ f_2(0), \ldots, c_{n-1} = F_{n-1}(0) = f_1 \circ f_2 \circ \cdots \circ f_{n-1}(0)\}$ consists of distinct points and such that the *cycle relation*

$$f_i \circ f_{i+1} \circ \cdots \circ f_{i+n-1}(0) = 0 \tag{13.7}$$

holds for all integers i. From this relation we see that $F_n(0) = 0$ and if $m = qn + r$, $0 \le r < n$, $F_m(0) = c_r$. The point here is that, because the iteration is backwards, we have to find successive pre-images of the points c_i.

Once we have done this, for any subsequence $F_{n_k} = f_1 \circ f_2 \circ \cdots \circ f_{n_k}$, we see that $F_{n_k}(0) \in S$ for all k. It follows that any limit function must map 0 to a point in S. Choosing subsequences appropriately, we can find n distinct limit functions G_i such that $G_i(0) = c_i$, $i = 0, \ldots, n-1$.

If X is Bloch, all limit functions must be constant so these are all the limit functions.

Suppose first that $n = 2$ and we are given two distinct points c_0 and c_1 in X. In this construction, all maps f_i will be different universal covering maps from Δ onto X. We may assume without loss of generality that $c_0 = 0$. We can find a universal covering map f_1 such that $f_1(0) = c_1$. Then because f_1 is defined up to a rotation about 0 and X is not relatively compact we can find an $x_1 \in X$ with $f_1(x_1) = 0$.

By the same reasoning we let f_2 be a universal covering map from Δ onto X such that $f_2(0) = x_1$ and such that there is an $x_2 \in X$ with $f_2(x_2) = 0$. Again there is such an x_2 because X is not relatively compact in Δ. Continuing this process we obtain a sequence of universal covering maps f_k and a sequence of points x_k in X such that

$$f_k(0) = x_{k-1} \text{ and } f_k(x_k) = 0 \qquad (13.8)$$

for all k. Choosing odd or even subsequences we obtain two distinct limit functions G_1, G_2 such that $G_1(0) = c_1$ and $G_2(0) = 0$.

For $n > 2$, the maps f_i are not universal covering maps. We need to apply Lemma 13.1.1 repeatedly. We carry out the proof for $n = 4$. It will then be clear how to carry it out for all n.

Let π_1 be a universal covering map such that $\pi_1(0) = c_1$. Choose $y_1, b_2', b_3' \in \Delta$ such that

$$\pi_1(y_1) = 0, \quad \pi_1(b_2') = c_2, \quad \pi_1(b_3') = c_3.$$

Use Lemma 13.1.1 to find h_1 and $x_1, b_2, b_3 \in X$ so that

$$h_1(x_1) = \frac{y_1}{x_1}, \quad h_1(b_2) = \frac{b_2'}{b_2}, \quad h_1(b_3) = \frac{b_3'}{b_3}.$$

Set $g_1(z) = z h_1(z)$; by construction it is a rational map. Then

$$
\begin{array}{ccccc}
      & x_1 & 0 & b_2 & b_3 \\
g_1 & \downarrow & \downarrow & \downarrow & \downarrow \\
      & y_1 & 0 & b_2' & b_3' \\
\pi_1 & \downarrow & \downarrow & \downarrow & \downarrow \\
      & 0 & c_1 & c_2 & c_3
\end{array}
$$

In the first and third lines the points are in X. In the second line the points are in Δ.

Set $f_1 = \pi_1 g_1$.

Choose a universal covering map π_2 such that $\pi_2(0) = b_2$ and find points y_{21}, y_2, b_{32}' in Δ so that

	y_{21}	y_2	0	b'_{32}
π_2	$\downarrow$	$\downarrow$	$\downarrow$	$\downarrow$
	x_1	0	b_2	b_3
g_1	$\downarrow$	$\downarrow$	$\downarrow$	$\downarrow$
	y_1	0	b'_2	b'_3
π_1	$\downarrow$	$\downarrow$	$\downarrow$	$\downarrow$
	0	c_1	c_2	c_3

Now use Lemma 13.1.1 again to find h_2 and points x_{21}, x_2, b_{32} in X so that, setting $g_2(z) = zh_2(z)$,

	x_{21}	x_2	0	b_{32}
g_2	$\downarrow$	$\downarrow$	$\downarrow$	$\downarrow$
	y_{21}	y_2	0	b'_{32}
π_2	$\downarrow$	$\downarrow$	$\downarrow$	$\downarrow$
	x_1	0	b_2	b_3
g_1	$\downarrow$	$\downarrow$	$\downarrow$	$\downarrow$
	y_1	0	b'_2	b'_3
π_1	$\downarrow$	$\downarrow$	$\downarrow$	$\downarrow$
	0	c_1	c_2	c_3

Set $f_2 = \pi_2 g_2$.

Again find a universal covering map π_3 and points y_{321}, y_{32}, y_3 in Δ and use Lemma 13.1.1 again to find h_3 and points x_{321}, x_{32}, x_3 in X such that, setting $g_3(z) = zh_3(z)$,

	x_{321}	x_{32}	x_3	0
g_3	$\downarrow$	$\downarrow$	$\downarrow$	$\downarrow$
	y_{321}	y_{32}	y_3	0
π_3	$\downarrow$	$\downarrow$	$\downarrow$	$\downarrow$
	x_{21}	x_2	0	b_{32}
g_2	$\downarrow$	$\downarrow$	$\downarrow$	$\downarrow$
	y_{21}	y_2	0	b'_{32}
π_2	$\downarrow$	$\downarrow$	$\downarrow$	$\downarrow$
	x_1	0	b_2	b_3
g_1	$\downarrow$	$\downarrow$	$\downarrow$	$\downarrow$
	y_1	0	b'_2	b'_3
π_1	$\downarrow$	$\downarrow$	$\downarrow$	$\downarrow$
	0	c_1	c_2	c_3

Set $f_3 = \pi_3 g_3$.

Finally, find a universal covering map π_4 and use the lemma to find $g_4 = zh_4$ so that

$$
\begin{array}{ccccc}
 & 0 & x_{432} & x_{43} & x_4 \\
g_4 & \downarrow & \downarrow & \downarrow & \downarrow \\
 & 0 & y_{432} & y_{43} & y_4 \\
\pi_4 & \downarrow & \downarrow & \downarrow & \downarrow \\
 & x_{321} & x_{32} & x_3 & 0 \\
g_3 & \downarrow & \downarrow & \downarrow & \downarrow \\
 & y_{321} & y_{32} & y_3 & 0 \\
\pi_3 & \downarrow & \downarrow & \downarrow & \downarrow \\
 & x_{21} & x_2 & 0 & b_{32} \\
g_2 & \downarrow & \downarrow & \downarrow & \downarrow \\
 & y_{21} & y_2 & 0 & b'_{32} \\
\pi_2 & \downarrow & \downarrow & \downarrow & \downarrow \\
 & x_1 & 0 & b_2 & b_3 \\
g_1 & \downarrow & \downarrow & \downarrow & \downarrow \\
 & y_1 & 0 & b'_2 & b'_3 \\
\pi_1 & \downarrow & \downarrow & \downarrow & \downarrow \\
 & 0 & c_1 & c_2 & c_3
\end{array}
$$

Set $f_4 = \pi_4 g_4$.

Note $f_1 \circ f_2 \circ f_3 \circ f_4(0) = 0$. This is the first cycle. We continue in this way to add new points to the cycles: to find f_5, f_6, f_7, we replace c_1, c_2 and c_3 with x_{432}, x_{43} and x_4, respectively; for the rest of the f_n, we repeat the process, changing the indices in the obvious way.

We find universal covering maps π_n and apply Lemma 13.1.1 to the pre-images in Δ under π_n of points in the incomplete cycles to get new points in X and maps h_n and g_n. The map $f_n = \pi_n g_n$ completes the next cycle and we have added points to the incomplete cycles. In the limit, for all $i \geq 1$ we have the cycle relation

$$f_i \circ f_{i+1} \circ f_{i+2} \circ f_{i+3}(0) = 0.$$

Let $F_k = f_1 \circ \cdots \circ f_k$. Then, $F_k(0) = c_r$ where $r = k \bmod 4$. The accumulation points are limits of subsequences $\{F_{n_k}\}$. For any such limit F, $F(0) = c_r$ for some $r = 0, \ldots, 3$. Because the c_r are distinct, we have at least four distinct accumulation points.

If X is Bloch, all the limit functions of this iterated function system are constant. Since $F(0) = c_k$ for some k, there are exactly four possible constant functions.

If $n \neq 4$, we have $n - 1$ steps before we complete the first cycle, but the process is the same. $\square$

We have

Corollary 13.1.1 *Let X be any non-relatively-compact subdomain of Δ and let L be any finite subset of X. Then there is an iterated function system whose accumulation points are exactly the constants in L.*

Proof. Suppose now that X is a non-relatively-compact subdomain of Δ and let $L = \{c_1, c_2, \ldots, c_n\}$ be a finite subset of X. Then by Exercise 10.12 there is a non-relatively-compact subdomain Y of X such that Y is Bloch in Δ and Y is still non-relatively-compact in Δ. Furthermore, by adjoining a relatively compact subdomain containing the points c_i if necessary, we may assume that Y contains all points c_i. Now Theorem 13.1.1 applied to the domain Y completes the proof. $\square$

13.2 Non-Bloch domains and non-constant limits

In this section we prove that a necessary condition for a subdomain X of Δ to be degenerate is that it be a Bloch subdomain. We also show that, for a non-Bloch subdomain X of Δ, any holomorphic function $f : \Delta \to X$ can be a limit function for some iterated function system. By the Riemann mapping theorem and Theorem 12.3.1 the analogous results hold for any simply connected hyperbolic domain Ω and subdomain X.

13.2.1 Preparatory lemmas

We begin with two preparatory lemmas. The first is a generalization of the fact that in a relatively compact subdomain of X the metrics ρ and ρ_X are equivalent. Basically it says that if distances are measured very deep inside X, where depth is measured in terms of the ρ-Bloch radius, the constant in the equivalence is close to 1 and depends on the depth. The proof is Exercise 7.5.

Lemma 13.2.1 *Let a be a point in a subdomain X of Δ. Let $C = R(X, \Delta, a)$ be the ρ-Bloch radius at a and assume $C > 1$. If z is a point in X such that $\rho(a, z) < 1$ then*
$$\rho_X(a, z) \leq (1 + \epsilon)\rho(a, z)$$
where $\epsilon = \epsilon(C) \to 0$ as $C \to \infty$.

208

Applications III
The second preparatory lemma is about the contraction properties of Blaschke product maps of degree 2. We know that Blaschke products are contractions in the disk and that the contraction constant is close to 1 near the boundary of Δ. Thus, if one pre-image of the origin is close to the boundary, the contraction constant near the origin is also close to 1. More precisely,

Lemma 13.2.2 *Let $c \neq 0$ be any point in Δ such that $\rho(0, c) < 1$. Set*

$$A_a(z) = \frac{z(z-a)}{1 - \overline{a}z}.$$

Then the pre-images $A_a^{-1}(c) = \{z_1, z_2\}$ may be labeled so that

$$\rho(0, z_1) = \rho(a, z_2) \to \rho(0, c) \text{ as } |a| \to 1.$$

The proof is Exercise 3.12.

13.2.2 A necessary condition for degeneracy

Theorem 13.2.1 *([31]) Suppose that $X \subset \Delta$ is not a Bloch subdomain of Δ. Then X is not degenerate in Δ.*

This theorem is a consequence of Theorem 13.2.2, but the direct proof below will make it easier to follow the proof of Theorem 13.2.2. In both proofs we use the estimates from the lemmas for the contraction constant of the Blaschke product near the origin and the equivalence of the metrics on Δ and X deep inside the non-Bloch domain to control the pre-images of two distinct points.

Proof of Theorem. Let X be any non-Bloch domain in Δ. We are going to construct an iterated function system from $\mathcal{H}ol(\Delta, X)$ with a non-constant accumulation point. Pick any two distinct points a_0 and w_0 in X such that $\rho_X(a_0, w_0) < 1/2$. We will recursively find functions f_n such that the iterated function system F_n will have a limit function F that satisfies $F(0) = a_0$ and $F(\tilde{w}) = w_0$ for some $\tilde{w}$, where $\rho(0, \tilde{w}) < 1$.

First let π_1 be a universal covering map from Δ onto X such that $\pi_1(0) = a_0$. Then there exists a point $c_0 \in \Delta$, such that $\pi_1(c_0) = w_0$ and

$$\rho(0, c_0) = \rho_X(a_0, w_0). \tag{13.9}$$

By Lemma 13.2.2, for any choice of $a_1 \in X$, the Blaschke product A_{a_1} produces two points w_1 and $\tilde{w}_1$ in Δ as the pre-images of c_0:

$$A_{a_1}(\tilde{w}_1) = A_{a_1}(w_1) = c_0 \text{ and } \rho(0, \tilde{w}_1) = \rho(a_1, w_1). \tag{13.10}$$

Define $f_1 = \pi_1 \circ A_{a_1}$; then

$$f_1(0) = f_1(a_1) = a_0, \tag{13.11}$$

$$f_1(w_1) = f_1(\tilde{w}_1) = w_0. \tag{13.12}$$

We need to make sure that the point w_1 belongs to X. To do this we use the preparatory lemmas.

Let $\epsilon_n \to 0$ be a sequence such that $\prod_1^\infty (1 + \epsilon_n)^2 \leq 2$. Since X is non-ρ-Bloch, by Lemma 13.2.2, we can choose $a_1 \in X$ with $|a_1|$ close enough to 1 that

$$\rho(0, \tilde{w}_1) = \rho(a_1, w_1) < (1 + \epsilon_1)\rho(0, c_0) = (1 + \epsilon_1)\rho_X(a_0, w_0). \tag{13.13}$$

Moreover, we may assume that $R = R(X, \Delta, a_1)$, the ρ-Bloch radius of the biggest disk centered at a_1 and contained in X, is greater than 1 that formula (13.13) implies that $w_1 \in X$. In fact, we may assume that R is so large that by Lemma 13.2.1 and (13.13) we have

$$\rho_X(a_1, w_1) < (1 + \epsilon_1)\rho(a_1, w_1) < (1 + \epsilon_1)^2 \rho_X(a_0, w_0) < 1. \tag{13.14}$$

Then, inductively, by our choice of ϵ_n, there exist points $a_n, w_n \in X$ and $\tilde{w}_n \in \Delta$ such that

$$f_n(0) = f_n(a_n) = a_{n-1} \text{ and } f_n(w_n) = f_n(\tilde{w}_n) = w_{n-1}, \tag{13.15}$$

$$\rho(a_n, w_n) = \rho(0, \tilde{w}_n) < (1 + \epsilon_n)\rho_X(a_{n-1}, w_{n-1}) \tag{13.16}$$

and

$$\rho_X(a_n, w_n) < (1 + \epsilon_n)\rho(a_n, w_n) < (1 + \epsilon_n)^2 \rho_X(a_{n-1}, w_{n-1}). \tag{13.17}$$

Therefore

$$\rho(0, \tilde{w}_n) < \prod_1^n (1 + \epsilon_i)\rho(0, c_0) < 1 \text{ and} \tag{13.18}$$

$$\rho_X(a_n, w_n) < \prod_1^n (1 + \epsilon_i)^2 \rho(0, c_0) < 1. \tag{13.19}$$

Now, if $F_n = f_1 \circ f_2 \circ f_3 \circ \cdots \circ f_n$, the formulas (13.15) yield

$$F_n(0) = a_0 \quad \text{and} \quad F_n(\tilde{w}_n) = w_0. \tag{13.20}$$

By Montel's theorem, F_n is a normal family so that any subsequence F_{n_j} converges locally uniformly to a holomorphic limit function F. Therefore, equations (13.20) yield

$$F(0) = a_0 \quad \text{and} \quad F(\tilde{w}) = w_0,$$

where $\tilde{w}$ is an accumulation point of the sequence $\tilde{w}_n$. Since, by formula (13.18 $\rho(0, \tilde{w}_n) < 1$ for all n, the point $\tilde{w}$ satisfies $\rho(0, \tilde{w}) \leq 1$ and so belongs to Δ. This implies that F is a non-constant function. $\square$

An even stronger theorem is the following.

Theorem 13.2.2 *([33]) If X is non-Bloch in Δ and $f : \Delta \to X$ is any holomorphic map, then f is the limit of an IFS.*

For the proof we use infinite Blaschke products instead of degree-2 Blaschke products.

Let $a_1, a_2, a_3, \ldots$ be a sequence of points in Δ and let k be a positive integer. The finite Blaschke products

$$A_n(z) = z^k \frac{z - a_1}{1 - \overline{a}_1 z} \frac{z - a_2}{1 - \overline{a}_2 z} \cdots \frac{z - a_n}{1 - \overline{a}_n z}$$

form a normal family. Thus, some subsequence $A_{n(l)}$ of A_n converges locally uniformly to an *infinite Blaschke product*

$$A(z) = z^k \frac{z - a_1}{1 - \overline{a}_1 z} \frac{z - a_2}{1 - \overline{a}_2 z} \cdots$$

If $|a_1 a_2 \ldots a_n| \to a > 0$, then $|A^{(k)}(0)| = k!a > 0$ for any accumulation point A and $A : \Delta \to \Delta$ is open.

As an infinite degree holomorphic self map of the disk, an infinite Blaschke product is a contraction. As for degree-2 products, though, the amount of contraction for points near the boundary of the disk is small. The following series of lemmas generalizes Lemma 13.2.2.

In these lemmas the notation $D(a, r)$ stands for a hyperbolic disk in Δ with center a and radius r.

Lemma 13.2.3 *Let $b_1, b_2, b_3, \ldots$ be a sequence of points in the unit disk Δ such that $|b_n| \to 1$. If $C > 1$ then there exists a subsequence $a_1, a_2, a_3, \ldots$ chosen from the b_n such that, for all $n \geq 1$ and all z in $D(a_n, \frac{n+4}{2})$, the finite Blaschke products*

$$A_n(z) = z^k \frac{z - a_1}{1 - \overline{a}_1 z} \frac{z - a_2}{1 - \overline{a}_2 z} \cdots \frac{z - a_n}{1 - \overline{a}_n z}$$

satisfy

$$|A_n(z)| \geq e^{\frac{-1}{2^{(n-2)}C}} \left| \frac{z - a_n}{1 - \overline{a}_n z} \right| |z|^{k-1}. \tag{13.21}$$

Proof. Assume first that $k = 1$. Pick a constant $C > 1$. Let $a_0 = 0$. Choose a_1 from the sequence $\{b_n\}$ such that $D(a_1, 4C)$ is disjoint from $D(0, C)$. Continue by choosing $a_2, a_3, \ldots$ such that the hyperbolic disks $D(a_n, 4^n C)$ are disjoint for all $n = 0, 1, 2, \ldots$ Let z be a point in $D(a_n, \frac{n+4}{2})$. Then, for $0 \le l < n$, we have

$$\rho(a_l, z) \ge \rho(a_l, a_n) - \rho(a_n, z) > (4^n + 4^l)C - \frac{n+4}{2} \ge 2^n 2^l C.$$

Since $\log(x) \le x$ for all positive x, for all l, $0 \le l < n$, we have

$$2^n 2^l C \le \rho(a_l, z) = \frac{1}{2} \log \frac{1 + \left|\frac{z - a_l}{1 - \bar{a}_l z}\right|}{1 - \left|\frac{z - a_l}{1 - \bar{a}_l z}\right|} \le \frac{1}{2} \log \frac{2}{1 - \left|\frac{z - a_l}{1 - \bar{a}_l z}\right|} \le \frac{1}{1 - \left|\frac{z - a_l}{1 - \bar{a}_l z}\right|}.$$

Therefore

$$\left|\frac{z - a_l}{1 - \bar{a}_l z}\right| \ge 1 - \frac{1}{2^n 2^l C}.$$

Note that

$$2x \ge \log \frac{1}{1 - x} \quad \text{whenever } 0 < x \le \frac{1}{2}.$$

Therefore,

$$\log \left|\frac{A_n(z)}{\frac{z - a_n}{1 - \bar{a}_n z}}\right| = \log |z| + \log \left|\frac{z - a_1}{1 - \bar{a}_1 z}\right| + \cdots + \log \left|\frac{z - a_{n-1}}{1 - \bar{a}_{n-1} z}\right|$$

$$\ge \sum_{l=0}^{n-1} \log \left(1 - \frac{1}{2^n 2^l C}\right) \ge -\sum_{l=0}^{n-1} \frac{1}{2^{n-1} 2^l C} \ge -\frac{1}{2^{(n-2)} C}.$$

This is equivalent to

$$|A_n(z)| \ge e^{-\frac{1}{2^{(n-2)} C}} \left|\frac{z - a_n}{1 - \bar{a}_n z}\right|$$

as claimed.

If $k > 1$ we can apply the above argument to $\tilde{A}_n(z) = A_n(z)/z^{k-1}$. Multiplying the result for $\tilde{A}_n$ through by $|z|^{k-1}$ we obtain the statement of the lemma. $\square$

We turn now to the infinite product $A(z)$.

Lemma 13.2.4 *Let A be an accumulation point of the finite Blaschke products $A_n(z)$ of Lemma 13.2.3. Then for all $n \geq 0$ and all $z \in D(a_n, \frac{n+4}{2})$ we have*

$$|A(z)| \geq e^{-\frac{1}{2^{(n-2)}C}} \left| \frac{z - a_n}{1 - \bar{a}_n z} \right| |z|^{k-1} \tag{13.22}$$

and the weaker inequality for $n \geq 1$,

$$|A(z)| \geq e^{-\frac{k}{2^{(n-2)}C}} \left| \frac{z - a_n}{1 - \bar{a}_n z} \right|. \tag{13.23}$$

Moreover, $A'(a_n) \neq 0$ for all $n \neq 0$ and $A(z)$ is univalent in a neighborhood of each a_n. In addition, if $k = 1$, $A'(0) \neq 0$ and A is univalent in a neighborhood of zero.

In particular, for $n = 0$ this says

$$|A(z)| \geq e^{-\frac{4}{C}} |z|^k \text{ whenever } \rho(0, z) < 2 \tag{13.24}$$

Proof of Lemma. We need to estimate the tail of the right side of

$$\log \left| \frac{A(z)}{\frac{z - a_n}{1 - \bar{a}_n z}} \right| = k \log |z| + \log \left| \frac{z - a_1}{1 - \bar{a}_1 z} \right| + \cdots$$

$$+ \log \left| \frac{z - a_{n-1}}{1 - \bar{a}_{n-1} z} \right| + \log \left| \frac{z - a_{n+1}}{1 - \bar{a}_{n+1} z} \right| + \cdots \tag{13.25}$$

(Notice that if $n = 0$ we replace k by $k - 1$.) To this end we note that for $l > n$ we have, with an argument similar to the one in Lemma 13.2.3,

$$\rho(a_l, z) \geq \rho(a_l, a_n) - \rho(a_n, z) > (4^l + 4^n)C - \frac{n+4}{2} \geq 2^n 2^l C$$

so that

$$2^n 2^l C \leq \rho(a_l, z) = \frac{1}{2} \log \frac{1 + \left| \frac{z - a_l}{1 - \bar{a}_l z} \right|}{1 - \left| \frac{z - a_l}{1 - \bar{a}_l z} \right|} \leq \frac{1}{2} \log \frac{2}{1 - \left| \frac{z - a_l}{1 - \bar{a}_l z} \right|} \leq \frac{1}{1 - \left| \frac{z - a_l}{1 - \bar{a}_l z} \right|}.$$

Therefore, for $l > n$,

$$\left| \frac{z - a_l}{1 - \bar{a}_l z} \right| \geq 1 - \frac{1}{2^n 2^l C}$$

and, again arguing as in Lemma 13.2.3,

$$\log \left| \frac{z - a_{n+1}}{1 - \bar{a}_{n+1} z} \right| + \cdots = \sum_{l=n+1}^{\infty} \log \left| \frac{z - a_l}{1 - \bar{a}_l z} \right| \geq \sum_{l=n}^{\infty} \frac{-1}{2^{n-1} 2^l C}.$$

This together with the proof of Lemma 13.2.3 implies

$$\log \left| \frac{A(z)}{\frac{z-a_n}{1-\bar{a}_n z}} \right| \geq \frac{-1}{2^{n-2}C} + (k-1)\log|z|$$

or, equivalently,

$$|A(z)| \geq e^{-\frac{1}{2^{(n-2)}C}} \left| \frac{z-a_n}{1-\bar{a}_n z} \right| |z|^{k-1}.$$

We obtain the second inequality by noting that

$$\log \left| \frac{A(z)}{\frac{z-a_n}{1-\bar{a}_n z}} \right| \geq k\log|z| + k\log \left| \frac{z-a_1}{1-\bar{a}_1 z} \right| + \cdots$$

$$+ k\log \left| \frac{z-a_{n-1}}{1-\bar{a}_{n-1} z} \right| + k\log \left| \frac{z-a_{n+1}}{1-\bar{a}_{n+1} z} \right| + \cdots \qquad (13.26)$$

Then from the arguments above we have

$$\log \left| \frac{A(z)}{\frac{z-a_n}{1-\bar{a}_n z}} \right| \geq k\sum_{l=0}^{\infty} \log \left(1 - \frac{1}{2^n 2^l C} \right) \geq -k\sum_{l=0}^{\infty} \frac{1}{2^{n-1} 2^l C} = -\frac{k}{2^{n-2}C}. \quad \square$$

The next step is an estimate for $k = 1$. It says that, for points close to the origin, A does not contract very much and is close to an isometry.

Lemma 13.2.5 *Suppose $k = 1$. For any $\epsilon > 0$, there exists a $C > 1$ such that the infinite Blaschke product A of Lemma 13.2.4 has the following property: for every point w in Δ with $\rho(0, w) < 1$, there exists a point z in Δ such that $A(z) = w$ and $\rho(0, z) \leq (1 + \epsilon)\rho(0, w)$.*

Proof. We claim that we may choose C sufficiently large that

$$\overline{D(0, 1)} \subset A(D(0, 2)). \qquad (13.27)$$

To prove this, let w be a point in $\overline{D(0, 1)} \setminus A(D(0, 2))$ closest to 0. Since $k = 1$, $A(0) = 0$ and A is univalent in a neighborhood of 0 so that $|w| > 0$. Thus, there exists a sequence $w_n = A(z_n)$ such that $\rho(0, z_n) < 2$ and $w_n \to w$. Passing to a subsequence we may assume $z_n \to z$ with $\rho(0, z) = 2$. The inequality (13.24) implies that $|w_n| \geq e^{-\frac{4}{C}}|z_n|$. Passing to the limit, we obtain $|w| \geq e^{-\frac{4}{C}}|z|$. Because $\rho(0, w) \leq 1$ and $\rho(0, z) = 2$, $|z| > |w|$. Now, if we choose C so large that $e^{-\frac{4}{C}}|z| > |w|$, we have a contradiction that proves the relation (13.27). Therefore, for every w in $D(0, 1)$ we conclude that there exists a point z in $D(0, 2)$ such that $A(z) = w$. The inequality (13.24) implies $\rho(0, z) \leq (1 + \epsilon)\rho(0, w)$ for the large C we have chosen. $\square$

When $k > 1$ we have

Lemma 13.2.6 *Suppose $k > 1$. Then any accumulation point $A(z)$ is an open map that satisfies*

$$\rho(0, A(z)) \le \rho(0, z^k).$$

Proof. The lemma follows by noting that $|A_n(z)| \le |z|^k$ for all n. $\square$

The last lemma of this group is

Lemma 13.2.7 *If p and q are any two points in Δ with $\rho(p, q) < 1$, then, given any $\sigma > 0$, for sufficiently large n there exist points p_n and q_n in $D(a_n, \frac{n+4}{2})$ such that $A(p_n) = p$, $A(q_n) = q$ and $\rho(p_n, q_n) \le (1+\sigma)\rho(p, q)$.*

Proof. Let p and q be any two points in Δ with $\rho(p, q) < 1$ and let $\sigma > 0$. Let $N \ge 2$ be the smallest integer n such that both p and q belong to $D(0, \frac{n}{8}+1)$. We claim that

$$\overline{D\left(0, \frac{n+4}{4}\right)} \subset A\left(D\left(a_n, \frac{n+4}{2}\right)\right) \tag{13.28}$$

for all sufficiently large n. To prove this claim, let $n \ge N$ and let w be a point in $\overline{D(0, \frac{n+4}{4})} \setminus A(D(a_n, \frac{n+4}{2}))$ closest to 0. Since $A(a_n) = 0$ we have $|w| > 0$. Thus, there exists a sequence $w_j = A(z_j)$ such that

$$z_j \in D\left(a_n, \frac{n+4}{2}\right), \rho(z_j, a_n) \to \frac{n+4}{2} \text{ and } w_j \to w.$$

Let $\tilde{z}_j \to \tilde{z}$ be a sequence such that $\rho(0, \tilde{z}_j) = \rho(a_n, z_j) \to \frac{n+4}{2}$. Inequality (13.23) implies that $|w_j| \ge e^{-\frac{k}{2^{(n-2)}c}}|\tilde{z}_j|$. Passing to the limit, we obtain $|w| \ge e^{-\frac{k}{2^{(n-2)}c}}|\tilde{z}|$ where $\rho(0, \tilde{z}) = \frac{n+4}{2}$. Therefore,

$$\frac{|w|}{|\tilde{z}|} < \frac{\frac{e^{\frac{n+4}{2}}-1}{e^{\frac{n+4}{2}}+1}}{\frac{e^{n+4}-1}{e^{n+4}+1}} < e^{-\frac{k}{2^{n-2}}},$$

which is a contradiction for all sufficiently large n. This proves the claim (13.28).

It now follows that there exist points p_n and q_n in $D(a_n, \frac{n+4}{2})$ such that $A(p_n) = p$ and $A(q_n) = q$. If $p = 0$, then we choose $p_n = a_n$ and the lemma follows directly from inequality (13.23). Thus, we may assume that both p and q are nonzero. By inequality (13.23), the points $\tilde{p}_n = (p_n - a_n)/(1 - \overline{a}_n p_n)$ and $\tilde{q}_n = (q_n - a_n)/(1 - \overline{a}_n q_n)$ are uniformly bounded away from both 0 and the unit circle. Furthermore, inequality (13.23) implies that the functions $\frac{1}{z}A(\frac{z+a_n}{1+\overline{a}_n z})$ converge locally uniformly to a point a on the unit circle as n approaches infinity. Therefore, $p/\tilde{p}_n$ and $q/\tilde{q}_n$ both converge to a, and, since $\rho(p_n, q_n) = \rho(\tilde{p}_n, \tilde{q}_n)$, the lemma follows. $\square$

13.2.3 Proof of Theorem 13.2.2

Proof. If X is a non-Bloch subdomain of Δ, there exists a sequence of points $a_1, a_2, a_3, \ldots$ in X such that the hyperbolic disks $D(a_n, n)$ with center at a_n and radius n satisfy

$$D(a_n, n) \subset X. \tag{13.29}$$

We will use these points to construct a sequence of infinite Blaschke products A_i, $i = 1, 2, \ldots$, that map X onto Δ and interleave them with covering maps π_i that map Δ onto X. The maps $f_i = \pi_{i-1} \circ A_i$, $i > 1$, will be maps in the iterated function system we need to construct. We will prove that the maps $G_n = A_1 \circ f_2 \circ \cdots \circ f_n$ converge to the identity. Setting $f_1 = f \circ A_1$ to complete the iterated function system proves the theorem.

Let z_0 be a point in the unit disk with $0 < \rho(0, z_0) < 1$. Choose a sequence $\epsilon_n > 0$ such that $1/\rho(0, z_0) > (1 + \epsilon_n)^{2n-1} \to 1$. Let $k = 1$. By Lemma 13.2.5, there exists a subsequence a_{n_j} of the a_n determining an infinite Blashke product

$$A_1(z) = z \frac{z - a_{n_1}}{1 - \overline{a}_{n_1} z} \frac{z - a_{n_2}}{1 - \overline{a}_{n_2} z} \frac{z - a_{n_3}}{1 - \overline{a}_{n_3} z} \cdots$$

and there is a point z_1 in Δ satisfying

$$A_1(z_1) = z_0 \text{ and } \rho(0, z_1) \leq (1 + \epsilon_1)\rho(0, z_0). \tag{13.30}$$

Now Lemma 13.2.7 with $p = 0$, $q = z_0$ and $\sigma = \epsilon_2$ implies that for any sufficiently large $n(1)$ there are points $p_1 = a_{n(1)}$ and q_1 in $D(a_{n(1)}, \frac{n(1)+4}{2})$ such that

$$A_1(p_1) = 0, \ A_1(q_1) = z_0 \text{ and } \rho(p_1, q_1) \leq (1 + \epsilon_2)\rho(0, z_0). \tag{13.31}$$

Observe that by inequality (13.23) A_1 maps X onto Δ.

Choose $n(1)$ so large that, by Lemma 13.2.1 and formula (13.29), we have

$$\rho_X(p_1, q_1) \leq (1 + \epsilon_2)\rho(p_1, q_1).$$

Next, choose a universal covering map π_1 from the unit disk Δ onto X such that $\pi_1(0) = p_1$ and $\pi_1(w_1) = q_1$ for some point w_1 with $\rho(0, w_1) = \rho_X(p_1, q_1)$.

Again by Lemma 13.2.5, there exists a subsequence of the a_{n_j} that determines another infinite Blaschke product $A_2(z)$ with $A_2(0) = 0$ and a point z_2 in Δ such that

$$A_2(z_2) = w_1 \text{ and } \rho(0, z_2) \leq (1 + \epsilon_2)\rho(0, w_1). \tag{13.32}$$

Thus

$$\rho(0, z_2) \leq (1 + \epsilon_2)^3 \rho(0, z_0).$$

Let $f_2 = \pi_1 \circ A_2$.

Now Lemmas 13.2.7 and 13.2.1 again, with $p = 0$, $q = z_0$ but $\sigma = \epsilon_3$, imply that for any sufficiently large $n(2)$ there are points $p_{21} = a_{n(2)}$ and q_{21} in $D(a_{n(2)}, \frac{n(2)+4}{2})$ such that

$$A_1(p_{21}) = 0, \ A_1(q_{21}) = z_0, \text{ and}$$

$$\rho_X(p_{21}, q_{21}) \leq (1 + \epsilon_3)\rho(p_{21}, q_{21}) \leq (1 + \epsilon_3)^2 \rho(0, z_0). \tag{13.33}$$

Since π_1 is a covering map, there exist points $\tilde{p}_{21}$ and $\tilde{q}_{21}$ such that $\pi_1(\tilde{p}_{21}) = p_{21}$, $\pi_1(\tilde{q}_{21}) = q_{21}$ and

$$\rho(\tilde{p}_{21}, \tilde{q}_{21}) = \rho_X(p_{21}, q_{21}) \leq (1 + \epsilon_3)^2 \rho(0, z_0).$$

Again choose $n(2) > n(1)$ sufficiently large and apply Lemmas 13.2.7 and 13.2.1, appropriately scaled, to the pair $(\tilde{p}_{21}, \tilde{q}_{21})$ to obtain points p_2 and q_2 such that $A_2(p_2) = \tilde{p}_{21}$, $A_2(q_2) = \tilde{q}_{21}$ and

$$\rho_X(p_2, q_2) \leq (1 + \epsilon_3)\rho(p_2, q_2) \leq (1 + \epsilon_3)^2 \rho(\tilde{p}_{21}, \tilde{q}_{21}) \leq (1 + \epsilon_3)^4 \rho(0, z_0).$$

Next, choose a universal covering map π_2 from the unit disk Δ onto X such that $\pi_2(0) = p_2$ and $\pi_2(w_2) = q_2$ for some point w_2 with $\rho(0, w_2) = \rho_X(p_2, q_2)$.

Repeating the arguments above, we find an infinite Blaschke product $A_3(z)$ with $A_3(0) = 0$ and a point z_3 in Δ such that

$$A_3(z_3) = w_2 \text{ and } \rho(0, z_3) \leq (1 + \epsilon_3)\rho(0, w_2).$$

Then

$$\rho(0, z_3) \leq (1 + \epsilon_3)^5 \rho(0, z_0).$$

Let $f_3 = \pi_2 \circ A_3$.

Continuing in this way we obtain a sequence of maps $f_n = \pi_{n-1} \circ A_n$ and a sequence of points z_n such that $\rho(0, z_n) \leq (1 + \epsilon_n)^{2n-1} \rho(0, z_0)$. Moreover, the maps

$$G_n = A_1 \circ f_2 \circ f_3 \circ \cdots \circ f_n$$

are holomorphic self maps of the unit disk that satisfy $G_n(0) = 0$ and $G_n(z_n) = z_0$. Therefore any accumulation point g of G_n is a holomorphic self map of the unit disk that maps an accumulation point $\tilde{z}$ of z_n to z_0. Thus, since $\rho(0, \tilde{z}) = \rho(0, z_0)$ the Schwarz lemma implies $g(z) = e^{i\theta}z$.

For any given $f \in \mathcal{H}ol(\Delta, X)$ set $f_1(z) = f(e^{-i\theta}A_1(z))$. Then

$$f_1 \circ f_2 \circ f_3 \circ \cdots = f(e^{-i\theta}A_1(f_2(f_3 \ldots)))$$

has f as an accumulation point. $\square$

Further results on limit functions are known. We state them here but direct the interested reader to the literature for proof. The following theorem and corollary, proved in [33], say that we can get any collection of maps at either end of the contraction spectrum; that is any collection of constants and hyperbolic isometries can occur as limit functions.

Theorem 13.2.3 *Let X be a non-Bloch subdomain of Δ. Let K be either an empty set or a compact subset of X and let C be any closed subset of $\overline{X}$. Let G be the set of covering maps $g \in \mathcal{H}ol(\Delta, X)$ such that $g(0) \in K$. Then there is a single iterated function system whose full set of accumulation points consists precisely of the open maps in G and the constants in C.*

Corollary 13.2.1 *Let X be a subdomain of the unit disk Δ. If X is a Bloch subdomain then every constant in X is an accumulation point of some iterated function system. On the other hand, if X is non-Bloch, then every holomorphic map from the unit disk to $\overline{X}$ is an accumulation point of some iterated function system.*

13.2.4 Equivalence of conditions

Theorems 13.2.1 and 13.2.2 show that the Bloch condition is a necessary condition for X to be degenerate in Δ and therefore all properties from levels I, II, III and IV are equivalent if the source domain Ω is the unit disk Δ. Moreover, Corollary 9.2.2 implies that c-Bloch subdomains of Δ are Bloch. Thus, all eight properties are equivalent. If we combine this result with Theorem 10.2.1 and Corollaries 10.4.1 and 10.4.2 we obtain

Corollary 13.2.2 *The following statements about a subdomain X of the unit disk Δ are equivalent.*

(a) X is degenerate in Δ.

(b) X is a Bloch domain.

(c) X is a Lipschitz domain.

(d) There exists $k < 1$ such that every holomorphic map f from Δ to X is a k-contraction with respect to the metric ρ_X. That is,

$$\rho_X(f(x), f(y)) \leq k\rho_X(x, y)$$

for all x and y in X.

(e) There exists $\tilde{k} < 1$ such that every holomorphic map f from Δ to X is a $\tilde{k}$-contraction with respect to the metric ρ. That is,

$$\rho(f(x), f(y)) \le \tilde{k}\rho(x, y)$$

for all x and y in Δ.

Therefore, when the source domain is the disk, the Lipschitz condition is the strongest condition one can give on the target domains of the iterated function system in order to guarantee that all accumulation functions be constant. In addition, the same condition is the strongest condition that guarantees that all maps be uniformly strong contractions. The equivalence of (d) and (e) does not hold in general as we saw in Section 10.5.

Exercise 13.1 (Open question) Are there a domain Ω and a degenerate subdomain X of Ω, such that X is not ck-Lip in Ω?

Exercise 13.2 (Open question) Are there a domain Ω and a subdomain X of Ω, such that X is k-Lip but not c-Lip in Ω?

Exercise 13.3 The equivalence of contraction properties (d) and (e) in Corollary 13.2.2 is a non-dynamical fact that is proved above using dynamics. D. Minda (personal communication) produced a relatively elementary direct non-dynamical proof of that equivalence. As an exercise try to do it yourself.

Exercise 13.4 Let X be a Bloch subdomain of Ω. Show that every holomorphic map f from Ω to X has a fixed point.

Exercise 13.5 *([9])* Let f be a holomorphic self map of the unit disk Δ with no fixed points. Show that the maps $f_n(z) = (1 - \frac{1}{n})f(z)$ do have fixed points z_n in the unit disk.

Exercise 13.6 With f and z_n as in the previous exercise:

1. Show that some subsequence of the hyperbolic disks with centers at z_n and radii $\rho(0, z_n)$ converge to a Euclidean disk $D \subset \Delta$ whose boundary contains the point 0 and some point t on the boundary of the unit disk.
2. Show that f maps D into itself and that the orbit $f^{\circ n}(x)$ converges to t for every point x in D.

14

Estimating hyperbolic densities

14.1 The smallest hyperbolic densities

Estimating the hyperbolic density for an arbitrary domain is an important task, because, except in a very few cases, there is no explicit formula for it. This is because the proof of the uniformization theorem is not constructive and thus there is no general formula for the universal covering map. The inclusion maps, however, provide one way to estimate the hyperbolic density of a given hyperbolic domain Ω by comparing it with those for which the formula is known. One such domain is a disk and we can use the formula from Chapter 2. If $d(z)$ is the Euclidean distance from z to the boundary of Ω, then the inclusion map from the Euclidean disk $D(\Omega, z)$ with center at z and radius $d(z)$ to Ω yields the estimate from above

$$\rho(z) \le \frac{1}{d(z)} \tag{14.1}$$

for any hyperbolic domain.

To obtain an estimate from below, we use the implicit formula for the hyperbolic density ρ_{ab} of $\mathbb{C} \setminus \{a, b\}$ which we derive from the formula in Theorem 14.2.3 for ρ_{01}. Then, if a and b are any two points in the complement of Ω, the inclusion map from Ω to $\mathbb{C} \setminus \{a, b\}$ yields

$$\sup \rho_{ab}(z) \le \rho_\Omega(z)$$

where the supremum is over all distinct pairs of points a and b in $\mathbb{C} \setminus \Omega$. The domains $\mathbb{C} \setminus \{a, b\}$ are the largest hyperbolic domains, and so carry the smallest hyperbolic densities.

14.2 A formula for ρ_{01}

In this section, we derive the formula for $\rho_{01}(z)$, the hyperbolic density of the plane punctured at the points $\{0, 1\}$. Computing the change under post-composition with a Möbius transformation, we obtain the formula for $\rho_{ab}(z)$, the hyperbolic density of the plane punctured at an arbitrary pair of points $\{a, b\}$. We call $\rho_{ab}(z)$ the *pair density*.

We need to find an explicit universal covering map from the upper half plane, $\mathbb{H} = \{\tau | \Im\tau > 0\}$, onto $\mathbb{C} \setminus \{0, 1\}$. To do this, we start with the Riemann map λ from the domain

$$W = \left\{\tau \,\middle|\, 0 < \Re\tau < 1 \text{ and } \left|\tau - \frac{1}{2}\right| > \frac{1}{2}\right\}$$

onto the upper half plane that satisfies

$$\lambda(0) = 1, \ \lambda(1) = \infty \text{ and } \lambda(\infty) = 0.$$

This map extends to the boundary of W by Theorem 3.4.2, and it takes on real values there.

Next, using the Schwarz reflection principle, we extend the map $\lambda(\tau)$ to the domain obtained by reflecting W across the imaginary axis

$$-\overline{W} = \{-\overline{\tau} | \tau \in W\}$$

so that

$$\lambda(\tau) = \overline{\lambda(-\overline{\tau})} \tag{14.2}$$

for all τ in $-\overline{W}$.

Similarly, we extend the map $\lambda(\tau)$ to the domain obtained by reflecting W across the vertical line $\tau = 1 + it, t > 0$,

$$2 - \overline{W} = \{2 - \overline{\tau} | \tau \in W\},$$

so that

$$\lambda(\tau) = \overline{\lambda(2 - \overline{\tau})} \tag{14.3}$$

for all τ in $2 - \overline{W}$.

We again reflect, this time across the third boundary $|\tau - 1/2| = 1/2$ so that

$$\lambda(\tau) = \overline{\lambda\left(\frac{1}{2}\left(1 + \frac{1}{2\overline{\tau} - 1}\right)\right)}$$

for all τ in $\frac{1}{2}(1 + \frac{1}{2\overline{W} - 1})$.

The extended λ is still real on the boundaries of these new domains. We continue extending λ by reflecting across boundaries and note that the images of W under the reflections fill up the entire upper half plane. The domain W is a fundamental domain for the group of reflections acting on $\mathbb{H}$ and, by Poincaré's theorem (see exercise 14.3 below), the domain $P = W \cup -\overline{W} \cup \{z | \Re z = 0\}$ is a fundamental domain for the group Γ acting on $\mathbb{H}$ generated by the side pairings

$$\tau \mapsto \tau + 2, \quad \tau \mapsto \frac{\tau}{2\tau + 1}.$$

By Exercise 14.1 below, the elements of Γ have the form

$$A = \begin{pmatrix} a & b \\ c & d \end{pmatrix}, \quad ad - bc = 1, \quad a, b, c, d \in \mathbb{Z},$$

where a and d are both odd and b and c are both even.

The images of P under Γ tile the upper half plane. Thus, if the function λ is extended to all of $\mathbb{H}$ by the rule $\lambda(\tau) = \lambda(A(\tau))$ where $A \in \Gamma$ and $A(\tau) \in P$, the extended function is a holomorphic map and its image is $\mathbb{C} \setminus \{0, 1\}$. It is also clear using the definition of the extension of λ that every path in $\mathbb{C} \setminus \{0, 1\}$ lifts to a path in the upper half plane and thus that λ is a universal covering map.

Therefore, formula (7.1) yields

$$\rho_{01}(\lambda(\tau))|\lambda'(\tau)| = \frac{1}{2\Im\tau}. \tag{14.4}$$

Equivalently,

$$\rho_{01}(z) = \frac{|g'z|}{2\Im g(z)} \tag{14.5}$$

where $g(z)$ is any inverse branch of $z = \lambda(\tau)$. Note that relations (14.2) and (14.4) yield one obvious property for the hyperbolic density ρ_{01}.

Theorem 14.2.1 *We have*

$$\rho_{01}(z) = \rho_{01}(\overline{z})$$

for all z in $\mathbb{C} \setminus \{0, 1\}$.

We have the following formula for $\lambda(\tau)$.

Theorem 14.2.2

$$\lambda(\tau) = 16q \prod_{n=1}^{\infty} \left(\frac{1 + q^{2n}}{1 + q^{2n-1}} \right)^8,$$

where $q = q(\tau) = e^{\pi i \tau}$.

Proof. A map $\lambda(\tau)$ is defined in the appendix as the cross ratio of the critical values of the Weierstrass $\wp$-function periodic with respect to the lattice generated by $(\pi, \pi 2)$.

Plugging the half periods $\pi/2$ and $(1 + \tau)\pi/2$ into equation (16.2) we obtain the formula of the theorem. This function $\lambda(\tau)$ also defines a holomorphic regular covering map from $\mathbb{H}$ onto $\mathbb{C} \setminus \{0, 1\}$ and is thus also a universal covering map. Any two universal covering maps differ by post-composition with a Möbius transformation leaving $\mathbb{C} \setminus \{0, 1\}$ invariant. The only possibilites are therefore

$$\lambda, \frac{1}{\lambda}, 1 - \lambda, \frac{1}{1 - \lambda}, \frac{\lambda - 1}{\lambda}, \frac{\lambda}{\lambda - 1}$$

and only the first fixes the boundary pointwise; it is thus the same as the function we just constructed.

In the appendix we show that, by using theta functions, $\lambda(\tau)$ can be expressed in terms of $q(\tau) = e^{\pi i \tau}$ as the infinite product above. $\quad\square$

We now use this formula to get an implicit formula for ρ_{01}. Taking the logarithms of both sides and differentiating yields

$$\log \lambda(\tau) = \log 16 + \log q + 8 \sum_{n=1}^{\infty} \log(1 + q^{2n}) - 8 \sum_{n=1}^{\infty} \log(1 + q^{2n-1}),$$

and

$$\frac{\lambda'(\tau)}{\lambda(\tau)} = \frac{q'}{q} \left(1 + 8 \sum_{n=1}^{\infty} \frac{2n q^{2n}}{1 + q^{2n}} - 8 \sum_{n=1}^{\infty} \frac{(2n-1) q^{2n-1}}{1 + q^{2n-1}} \right). \tag{14.6}$$

Noting that $\frac{q'}{q} = i\pi$ and substituting into equation (14.4) we obtain the following implicit formula for ρ_{01}.

Theorem 14.2.3

$$\rho_{01} \left(16 q \prod_{n=1}^{\infty} (\frac{1 + q^{2n}}{1 + q^{2n-1}})^8 \right)$$

$$= \frac{1}{32 \pi |q| \prod_{n=1}^{\infty} |\frac{1 + q^{2n}}{1 + q^{2n-1}}|^8 |1 + 8 \sum_{n=1}^{\infty} \frac{2n q^{2n}}{1 + q^{2n}} - 8 \sum_{n=1}^{\infty} \frac{(2n-1) q^{2n-1}}{1 + q^{2n-1}}| \Im \tau}$$

where $q = q(\tau) = e^{\pi i \tau}$.

We may apply an affine transformation that sends the points $0, 1$ respectively to any distinct points a, b to obtain the formula for the hyperbolic density ρ_{ab} of $\mathbb{C} \setminus \{a, b\}$. We have $M(z) = (b - a)z + a$ so that

$$\rho_{ab}(M(\lambda)) = \frac{\rho_{01}(\lambda)}{|M'(\lambda)|} = \frac{\rho_{01}(\lambda)}{|b-a|}. \tag{14.7}$$

Exercise 14.1 Let Γ be the group generated by the Möbius transformations

$$\tau \mapsto \tau + 2, \quad \tau \mapsto \frac{\tau}{2\tau + 1}.$$

Show that elements of Γ have the matrix form

$$A = \begin{pmatrix} a & b \\ c & d \end{pmatrix}, \quad ad - bc = 1, \quad a, b, c, d \in \mathbb{Z},$$

where a and d are both odd and b and c are both even.

Exercise 14.2 1. Use the symmetry of P under the map $\tau \mapsto -1/\tau$ to show that $\lambda(i) = 1/2$.

2. Show that $\lambda(1+i) = -1$.

3. Use formula (14.6) to show that $|\frac{\lambda'(1+i)}{\lambda(1+i)}| \approx 0.7$ and then use Theorem 14.2.3 to conclude that $\rho_{01}(-1) \approx .114$.

Exercise 14.3 Show that the domain P defined in this section satisfies the conditions of Poincaré's theorem and is therefore a fundamental domain for the group Γ.

14.3 A lower bound on ρ_{01}

In order to find a lower bound for $\rho_{01}(\lambda)$ we need to get control of its local behavior. We show here how to do this. More details can be found in [22]. To do this we use Theorem 14.2.2 to obtain the expansion

$$\lambda(\tau) = 16q - 128q^2 + \cdots,$$

clearly valid for $\Im \tau$ large. Inverting we obtain the expansion

$$e^{\pi i g(z)} = q(z) = \frac{1}{16}z + \frac{1}{32}z^2 + \cdots, \tag{14.8}$$

where $\tau = g(z)$ is any inverse branch of $z = \lambda(\tau)$ which is valid for z close to zero.

We write $z = x + iy = re^{i\theta} = e^{\sigma + i\theta}$ where $\sigma(z) = \log r$ and set $u(z) = \log \rho_{01}(z)$. We find estimates for the partial derivatives u_σ and u_θ in neighborhoods of the punctures $\{0, 1, \infty\}$ and use them to estimate the asymptotics of ρ_{01} in these neighborhoods. We then investigate the partial derivatives $\partial \rho_{01}/\partial \theta$, $\partial \rho_{01}/\partial r$ and $\partial \rho_{01}/\partial y$ in these neighborhoods. Note that because of the subscripts we have switched notation for the partial derivatives. Finally, we use these results to obtain an explicit lower bound on the growth of ρ_{01} near the punctures.

14.3.1 The first estimates

We use the formulas we developed in Section 14.2 to get estimates on the derivatives of $\log \rho_{01}(\lambda)$ in a neighborhood of the puncture at the origin. We then use the transformations $\lambda \mapsto 1 - \lambda$ and $\lambda \mapsto 1/\lambda$ to get estimates on these derivatives near 1 and infinity. The reader is referred to [64] for more details.

From equation (14.5) we have $u(z) = \log \rho_{01}(z) = \log |g'(z)| - \log \Im g(z) - \log 2$. On $\mathbb{C} \setminus \{0, 1\}$ define

$$R(z) = \frac{\partial u}{\partial \sigma}(z) = u_z(z) z_\sigma(z) + u_{\bar{z}}(z) \bar{z}_\sigma(z)$$

and

$$S(z) = \frac{\partial u}{\partial \theta}(z) = u_z(z) z_\theta(z) + u_{\bar{z}}(z) \bar{z}_\theta(z).$$

Therefore,

$$R(z) = \frac{\partial \log |g'(z)|}{\partial \sigma} - \frac{\partial \log \Im g(z)}{\partial \sigma},$$

and

$$S(z) = \frac{\partial \log |g'(z)|}{\partial \theta} - \frac{\partial \log \Im g(z)}{\partial \theta}.$$

A simple calculation shows

$$R(z) = \Re \left(\frac{z g''(z)}{g'(z)} \right) - \frac{\Im z g'(z)}{\Im g(z)} \tag{14.9}$$

and

$$S(z) = -\Im \left(\frac{z g''(z)}{g'(z)} \right) - \frac{\Re z g'(z)}{\Im g(z)}. \tag{14.10}$$

The next lemma estimates $R(z)$ and $S(z)$ near the origin.

Lemma 14.3.1 *For $z = x + iy = re^{i\theta}$ and r close to zero set*

$$\delta(r) = \frac{1}{\log 16 - \log r}.$$

Then

$$R(z) = -1 + \delta(r) + \frac{x(1 + \delta(r))}{2} + o(r\delta(r)),$$

and

$$S(z) = -\frac{y(1+\delta(r))}{2} + o(r\delta(r)),$$

so that

$$\lim_{z\to 0} R(z) = -1 \qquad\qquad (14.11)$$

and

$$\lim_{z\to 0} S(z) = 0. \qquad\qquad (14.12)$$

Proof. Formula (14.8) yields

$$\Im g(z) = -\frac{\log|q(z)|}{\pi} = \frac{1}{\pi\delta(r)} - \frac{\Re z}{2\pi} + O(r^2), \qquad (14.13)$$

$$\pi i g'(z) = \frac{q'(z)}{q(z)} = \frac{1}{z} + \frac{1}{2} + O(z), \qquad (14.14)$$

and

$$z\frac{g''(z)}{g'(z)} = z\left(\frac{q''(z)}{q'(z)} - \frac{q'(z)}{q(z)}\right) = -1 + \frac{z}{2} + O(r^2).$$

This together with (14.9) implies

$$R(z) = -1 + \frac{x}{2} + \delta(r)\frac{1+\frac{x}{2}}{1-\frac{\delta(r)x}{2}} + o(r\delta(r)).$$

Thus

$$R(z) = -1 + \frac{x}{2} + \delta(r) + \frac{\delta(r)x}{2} + o(r\delta(r)),$$

so that

$$R(z) = -1 + \delta(r) + \frac{x(1+\delta(r))}{2} + o(r\delta(r)).$$

On the other hand, (14.10) implies

$$S(z) = -\frac{y}{2} - \delta(r)\frac{\frac{y}{2}}{1-\frac{\delta(r)x}{2}} + o(r\delta(r)).$$

Thus

$$S(z) = -\frac{y}{2} - \frac{\delta(r)y}{2} + o(r\delta(r)),$$

and

$$S(z) = -\frac{y(1+\delta(r))}{2} + o(r\delta(r)). \quad \square$$

The transformation $z \longmapsto 1 - z$ preserves $\mathbb{C} - \{0, 1\}$ and interchanges 0 and 1. In the following lemma we use this fact, together with the previous estimate, to get estimates of $R(z)$ and $S(z)$ for r close to 1.

Lemma 14.3.2 *For $z = x + iy = re^{i\theta}$ and r close to 1 we have*

$$R(z) = -1 + \frac{1 - r^2}{2|1 - z|^2}(1 - \delta(|1 - z|)) + O(\delta(|1 - z|))$$

and

$$S(z) = -\frac{y}{|1 - z|^2}(1 - \delta(|1 - z|)) + o(\delta(|1 - z|)).$$

Proof. Since $\delta(r) = 1/(\log 16 - \log r)$ we have $r = o(\delta(r))$. Thus, applying Lemma 14.3.1, we obtain $R(1 - z) = -1 + O(\delta(|1 - z|))$ and $S(1 - z) = o(\delta(|1 - z|))$.

To get the estimates on $R(z)$ and $S(z)$ in this neighborhood we apply the chain rule. For readability in the following formulas involving partial derivatives, we will temporarily use the notation ρ^{01} for ρ_{01}. By the chain rule we have

$$R(z) = \frac{\rho_z^{01}(z)}{\rho^{01}(z)} z + \frac{\rho_{\bar{z}}^{01}(z)}{\rho^{01}(z)} \bar{z} \tag{14.15}$$

and

$$S(z) = \frac{\rho_z^{01}(z)}{\rho^{01}(z)} iz - \frac{\rho_{\bar{z}}^{01}(z)}{\rho^{01}(z)} i\bar{z}. \tag{14.16}$$

The conformal map $z \mapsto 1 - z$ preserves $\overline{\mathbb{C}} \setminus \{0, 1, \infty\}$, and so we have

$$\rho^{01}(z) = \rho^{01}(1 - z). \tag{14.17}$$

Thus, after taking partial derivatives,

$$\rho_z^{01}(z) = -\rho_z^{01}(1 - z)$$

and

$$\rho_{\bar{z}}^{01}(z) = -\rho_{\bar{z}}^{01}(1 - z).$$

Therefore (14.15) and (14.16) imply

$$R(1 - z) = \frac{\rho_z^{01}(z)}{\rho^{01}(z)}(z - 1) - \frac{\rho_{\bar{z}}^{01}(z)}{\rho^{01}(z)}(1 - \bar{z}) \tag{14.18}$$

and

$$S(1-z) = \frac{\rho_z^{01}(z)}{\rho^{01}(z)}i(z-1) + \frac{\rho_{\bar{z}}^{01}(z)}{\rho^{01}(z)}i(1-\bar{z}). \qquad (14.19)$$

Combining (14.18) and (14.19), we easily find

$$\frac{\rho_{\bar{z}}^{01}(z)}{\rho^{01}(z)} = \frac{-R(1-z) - iS(1-z)}{2(1-\bar{z})} \qquad (14.20)$$

and

$$\frac{\rho_z^{01}(z)}{\rho^{01}(z)} = \frac{-R(1-z) + iS(1-z)}{2(1-z)}. \qquad (14.21)$$

Combining (14.15) and (14.18), we find

$$R(z) - R(1-z) = \frac{\rho_z^{01}(z)}{\rho^{01}(z)} + \frac{\rho_{\bar{z}}^{01}(z)}{\rho^{01}(z)}. \qquad (14.22)$$

Similarly, combining (14.16) and (14.19), we find

$$S(z) - S(1-z) = \frac{\rho_z^{01}(z)}{\rho^{01}(z)}i - \frac{\rho_{\bar{z}}^{01}(z)}{\rho^{01}(z)}i. \qquad (14.23)$$

Thus

$$R(z) = R(1-z) - R(1-z)\frac{1-x}{|1-z|^2} - S(1-z)\frac{y}{|1-z|^2}$$

and

$$S(z) = S(1-z) + R(1-z)\frac{y}{|1-z|^2} - S(1-z)\frac{1-x}{|1-z|^2}.$$

Next, using the estimates in Lemma 14.3.1, we get for $R(z)$

$$R(z) = -1 - \left(-1 + \delta(|1-z|) + \frac{(1-x)(1+\delta(|1-z|))}{2}\right)\frac{1-x}{|1-z|^2}$$
$$-\frac{y^2(1+\delta(|1-z|))}{2|1-z|^2} + O(\delta(|1-z|)).$$

Rearranging, we have

$$R(z) = -1 + \frac{(1-\delta(|1-z|))(1-x)}{|1-z|^2} - \frac{(1+\delta(|1-z|))(1-x)^2}{2|1-z|^2}$$
$$-\frac{y^2(1+\delta(|1-z|))}{2|1-z|^2} + O(\delta(|1-z|)).$$

Collecting terms we get

$$R(z) = -1 + \frac{(1-\delta(|1-z|))(1-x)}{|1-z|^2} - \frac{1+\delta(|1-z|)}{2} + O(\delta(|1-z|)),$$

and, absorbing $-\delta(|1-z|)$ into the $O(\delta(|1-z|))$ term,

$$R(z) = -1 + \frac{(1-\delta(|1-z|))(1-x)}{|1-z|^2} - \frac{1-\delta(|1-z|)}{2} + O(\delta(|1-z|)).$$

Thus,

$$R(z) = -1 + \frac{1-r^2}{2|1-z|^2}(1-\delta(|1-z|)) + O(\delta(|1-z|)).$$

Doing the same for $S(z)$ we get

$$S(z) = S(1-z) + \left(-1 + \delta(|1-z|) + \frac{(1-x)(1+\delta(|1-z|))}{2} \right.$$
$$+ o(|1-z|\delta(|1-z|)) \Bigg) \frac{y}{|1-z|^2}$$
$$- \left(\frac{y(1+\delta(|1-z|))}{2} + o(|1-z|\delta(|1-z|)) \right) \frac{1-x}{|1-z|^2},$$

$$S(z) = S(1-z) - \frac{y}{|1-z|^2}(1-\delta(|1-z|)) + \frac{(1-x)y(1+\delta(|1-z|))}{2|1-z|^2}$$
$$+ o\left(\frac{y\delta(|1-z|)}{|1-z|} \right) - \frac{(1-x)y(1+\delta(|1-z|))}{2|1-z|^2} + o\left(\frac{(1-x)\delta(|1-z|)}{|1-z|} \right),$$

$$S(z) = S(1-z) - \frac{y}{|1-z|^2}(1-\delta(|1-z|)) + o(\delta(|1-z|)).$$

Finally,

$$S(z) = -\frac{y}{|1-z|^2}(1-\delta(|1-z|)) + o(\delta(|1-z|)). \quad \square$$

To obtain estimates for $R(z)$ and $S(z)$ in a neighborhood of infinity, we use the transformation $z \longmapsto 1/z$ which preserves $\overline{\mathbb{C}} - \{0, 1, \infty\}$ and interchanges 0 and ∞, and then apply Lemma 14.3.1.

Lemma 14.3.3 *For $z = x + iy = re^{i\theta}$ in a neighborhood of infinity we have*

$$R(z) = -1 - \delta\left(\frac{1}{r} \right) - \frac{x(1+\delta(\frac{1}{r}))}{2r^2} + o\left(\frac{1}{r}\delta\left(\frac{1}{r} \right) \right)$$

and

$$S(z) = -\frac{y(1+\delta(\frac{1}{r}))}{2r^2} + o\left(\frac{1}{r}\delta\left(\frac{1}{r} \right) \right).$$

Proof. The transformation $z \mapsto \frac{1}{z}$ preserves $\overline{\mathbb{C}} \setminus \{0, 1, \infty\}$, and so

$$\rho_{01}(z) = \frac{1}{|z|^2} \rho_{01}\left(\frac{1}{z}\right). \tag{14.24}$$

Therefore,

$$u(z) = u(e^\sigma e^{i\theta}) = \log \rho_{01}(e^\sigma e^{i\theta}) = -2\sigma + \log \rho_{01}(e^{-\sigma} e^{-i\theta}).$$

Thus, after taking derivatives,

$$R(z) = -R\left(\frac{1}{z}\right) - 2,$$

and

$$S(z) = -S\left(\frac{1}{z}\right).$$

Lemma 14.3.1 immediately yields

$$R(z) = -1 - \delta\left(\frac{1}{r}\right) - \frac{x(1 + \delta(\frac{1}{r}))}{2r^2} + o\left(\frac{1}{r}\delta\left(\frac{1}{r}\right)\right),$$

and

$$S(z) = -\frac{y(1 + \delta(\frac{1}{r}))}{2r^2} + o\left(\frac{1}{r}\delta\left(\frac{1}{r}\right)\right). \quad \square$$

14.3.2 Estimates of ρ_{01} near the punctures

Now that we have estimates of the partial derivatives of $\log \rho_{01}(z)$ we study the behavior of the hyperbolic density $\rho_{01}(z)$ itself near the three boundary points.

Lemma 14.3.4 *We have*

$$\rho_{01}(z) = \frac{\delta(|z|)}{2|z|} + \frac{x\delta(|z|)}{4|z|} + o(\delta(|z|)) \ near \ z = 0,$$

$$\rho_{01}(z) = \frac{\delta(|1-z|)}{2|1-z|} + \frac{(1-x)\delta(|1-z|)}{4|1-z|} + o(\delta(|1-z|)) \ near \ z = 1,$$

and

$$\rho_{01}(z) = \frac{\delta(\frac{1}{|z|})}{2|z|} + \frac{x\delta(\frac{1}{|z|})}{4|z|^3} + o\left(\frac{\delta(\frac{1}{|z|})}{|z|^2}\right) \ near \ z = \infty.$$

Proof. First assume that $z = x + iy$ is near 0. Combining (14.5), (14.13) and (14.14) we get

$$\rho_{01}(z) = \frac{|\frac{1}{z} + \frac{1}{2} + O(|z|)|}{\frac{2}{\delta(|z|)} - x + O(|z|^2)}.$$

Therefore

$$\rho_{01}(z) = \frac{\delta(|z|)}{2|z|} \left| 1 + \frac{z}{2} + O(|z|^2) \right| \left(1 + \frac{\delta(|z|)x}{2} + O(\delta(|z|)|z|^2) \right),$$

and working with the $|1 + \frac{z}{2} + O(|z|^2)|$ term we get

$$\rho_{01}(z) = \frac{\delta(|z|)}{2|z|} \sqrt{1 + x + O(|z|^2)} \left(1 + \frac{\delta(|z|)x}{2} + O(\delta(|z|)|z|^2) \right),$$

and then

$$\rho_{01}(z) = \frac{\delta(|z|)}{2|z|} \left(1 + \frac{x}{2} + O(|z|^2) \right) \left(1 + \frac{\delta(|z|)x}{2} + O(\delta(|z|)|z|^2) \right).$$

Expanding and collecting terms we have

$$\rho_{01}(z) = \frac{\delta(|z|)}{2|z|} \left(1 + \frac{x(1 + \delta(|z|))}{2} + O(|z|^2) \right),$$

and finally

$$\rho_{01}(z) = \frac{\delta(|z|)}{2|z|} \left(1 + \frac{x}{2} + o(|z|) \right),$$

proving the first formula of the lemma.

The other two follow by applying (14.17) and (14.24). In particular, near $z = x + iy = \infty$, we have

$$\rho_{01}(z) = \frac{1}{|z|^2} \rho_{01}\left(\frac{1}{z} \right)$$

$$= \frac{1}{|z|^2} \left(\frac{\delta(\frac{1}{|z|})}{2\frac{1}{|z|}} + \frac{x\delta(\frac{1}{|z|})}{4|z|^2 \frac{1}{|z|}} + o\left(\delta\left(\frac{1}{|z|} \right) \right) \right).$$

A similar simple calculation gives the formula for the estimate near 1. □

14.3.3 The derivatives of ρ_{01}

Now we study how the hyperbolic density $\rho_{01}(z)$ changes, where $z = re^{i\theta} = x + iy$, and we vary θ, r or y. For this, we need the following lemma.

Lemma 14.3.5 *Let $p(z)$ be a positive function defined on a plane domain Ω. If a C^2 function $\lambda(z)$ defined in Ω satisfies*

$$\Delta\lambda(z) = p(z)\lambda(z) \quad \text{for all } z \in \Omega \tag{14.25}$$

and if

$$\liminf_{z \to w} \lambda(z) \geq 0 \tag{14.26}$$

for every boundary point (including possibly ∞) of Ω, then

$$\lambda(z) \geq 0 \text{ for all } z \in \Omega.$$

Proof. Suppose $0 \in \Omega$ and $\lambda(0) < 0$. It follows that the global infimum of $\lambda(z)$ over all z in Ω is negative and by the second hypothesis, (14.26), that it is achieved for some point $z_0 \in \Omega$. Since the Laplacian of a function is non-negative at a minimum, the first hypothesis (14.25) cannot hold. $\square$

First we vary θ and then we vary r. We have

Proposition 14.3.1 *The following hold:*

$$\frac{\partial \rho_{01}}{\partial \theta} \geq 0 \text{ for } -\pi < \theta < 0; \tag{14.27}$$

$$\frac{\partial \rho_{01}}{\partial \theta} \leq 0 \text{ for } 0 < \theta < \pi; \tag{14.28}$$

$$\frac{\partial \rho_{01}}{\partial r} \leq -\frac{\rho_{01}}{r} \text{ for } r > 1; \tag{14.29}$$

$$\frac{\partial \rho_{01}}{\partial r} \geq -\frac{\rho_{01}}{r} \text{ for } 0 < r < 1. \tag{14.30}$$

Proof. In Section 7.5 we saw that ρ_{01} has constant curvature equal to -4 and that

$$\Delta \log \rho_{01}(z) = 4\rho_{01}^2(z).$$

If we put $u = \log \rho_{01}$ into this equation we get

$$\Delta u = 4e^{2u}.$$

Furthermore,

$$u_\sigma = u_x e^\sigma \cos \theta + u_y e^\sigma \sin \theta,$$

$$u_{\sigma\sigma} = u_{xx} e^{2\sigma} \cos^2 \theta + u_{yy} e^{2\sigma} \sin^2 \theta + u_x e^\sigma \cos \theta + u_y e^\sigma \sin \theta,$$

$$+ 2u_{xy} e^{2\sigma} \sin \theta \cos \theta$$

and

$$u_\theta = -u_x e^\sigma \sin\theta + u_y e^\sigma \cos\theta,$$

$$u_{\theta\theta} = u_{xx} e^{2\sigma} \sin^2\theta + u_{yy} e^{2\sigma} \cos^2\theta - u_x e^\sigma \cos\theta - u_y e^\sigma \sin\theta.$$

$$- 2u_{xy} e^{2\sigma} \sin\theta \cos\theta$$

Thus,

$$u_{\sigma\sigma} + u_{\theta\theta} = \Delta u e^{2\sigma} = 4e^{2(\sigma+u)}. \tag{14.31}$$

Since $S = u_\theta$ we can rewrite this as

$$S_{\sigma\sigma} + S_{\theta\theta} = (4e^{2(\sigma+u)})_\theta = 8Se^{2(\sigma+u)}. \tag{14.32}$$

Similarly,

$$R_{\sigma\sigma} + R_{\theta\theta} = (4e^{2(\sigma+u)})_\sigma = 8(1+R)e^{2(\sigma+u)}. \tag{14.33}$$

Now we apply Lemma 14.3.5 with Ω equal to the strip $\{(\sigma, \theta) \mid 0 < \theta < \pi\}$, $\lambda = -S$ and the function $p(z)$ defined by $p(z) = 8e^{2(\sigma+u)} = 8(|z|\rho_{01}(z))^2 \geq 0$. By equation (14.32) we see that hypothesis (14.25) holds. The symmetry of ρ_{01} with respect to the real axis in Theorem 14.2.1 implies the symmetry

$$S(z) = -S(\bar{z}) \tag{14.34}$$

for all z in $\mathbb{C} \setminus \{0, 1\}$. Therefore $S(x) = 0$ for all real x different from 0 or 1.

We see from Lemmas 14.3.1, 14.3.2 and 14.3.3 that the limsup as S tends to each of the boundary points is non-positive so that hypothesis (14.26) of Lemma 14.3.5 holds and we conclude that $S(z) \leq 0$ in the upper half plane. This, together with equation (14.34), implies $S(z) \geq 0$ in the lower half plane. Thus, we have inequalities (14.27) and (14.28).

Similarly, we apply Lemma 14.3.5 with Ω equal to the strip $\{(\sigma, \theta) \mid \sigma < 0\}$, $\lambda(z) = R(z) + 1$ and the function $p(z) = 8e^{2(\sigma+u)} = 8(|z|\rho_{01}(z))^2 \geq 0$. By equation (14.33) we see that hypothesis (14.25) holds. Equation (14.24) and Theorem 14.2.1 imply

$$u(z) = u\left(\frac{1}{\bar{z}}\right) - 2\sigma,$$

and differentiation easily produces

$$R(z) = -2 - R\left(\frac{1}{\bar{z}}\right) \tag{14.35}$$

for all z in $\mathbb{C} \setminus \{0, 1\}$. Therefore, for all $z \neq 1$ on the unit circle, $R(z) = -1$. We see from Lemmas 14.3.1 and 14.3.2 that the liminf of R as z tends to each of the boundary points is not less than -1 so that hypothesis (14.26) of Lemma 14.3.5 holds.

Therefore by Lemma 14.3.5 we conclude that

$$R(z) \geq -1 \tag{14.36}$$

in the punctured unit disk Δ^*. This, together with the relation (14.35), implies

$$R(z) \leq -1 \tag{14.37}$$

in the exterior of the closed unit disk. Since

$$\frac{\partial \rho_{01}}{\partial r} = \frac{\rho_{01}}{r} u_\sigma,$$

we obtain inequalities (14.29) and (14.30). $\square$

Now we vary y.

Proposition 14.3.2 *We have*

$$\frac{\partial \rho_{01}}{\partial y} \leq 0 \text{ for } y > 0, \tag{14.38}$$

$$\frac{\partial \rho_{01}}{\partial y} \geq 0 \text{ for } y < 0. \tag{14.39}$$

Proof. Let

$$Q(z) = \frac{\partial u(z)}{\partial y}.$$

The symmetry in Theorem 14.2.1 implies

$$Q(z) = -Q(\bar{z}) \tag{14.40}$$

for all z in $\mathbb{C} \setminus \{0, 1\}$ so that $Q(x) = 0$ for all real x different from 0 and 1. Furthermore

$$Q(z) = \frac{\partial u}{\partial \sigma} \frac{\partial \sigma}{\partial y} + \frac{\partial u}{\partial \theta} \frac{\partial \theta}{\partial y},$$

or

$$Q(z) = R(z) \frac{y}{r^2} + S(z) \frac{x}{r^2}. \tag{14.41}$$

Using the estimates in Lemma 14.3.1 and collecting terms, we see that, for z near 0,

$$Q(z) = -\frac{y}{r^2}(1 - \delta(r)) + o(\delta(r)). \tag{14.42}$$

Let Ω be the upper half plane, and let $\lambda(z) = -Q(z)$. Then $\liminf \lambda(z_n) \geq 0$ for every sequence z_n in Ω converging to 0.

The symmetry of formula (14.17) implies

$$Q(z) = -Q(1-z).$$

Using (14.42) we see that, for z near 1,

$$Q(z) = -\frac{y(1-\delta(|1-z|))}{|1-z|^2} + o(\delta(|1-z|)).$$

Thus $\liminf \lambda(z_n) \geq 0$ for every sequence z_n in the upper half plane converging to 1.

Finally, when z is near infinity, using the estimates in Lemma 14.3.3 and the relation (14.41) and collecting terms, we see that near ∞

$$Q(z) = -\frac{xy(1+\delta(\frac{1}{r}))}{r^4} - \frac{y}{r^2}\left(1+\delta\left(\frac{1}{r}\right)\right) + o\left(\frac{1}{r^2}\delta\left(\frac{1}{r}\right)\right).$$

Thus $\liminf \lambda(z_n) \geq 0$ for every sequence z_n in the upper half plane converging to ∞. Therefore, $\lambda(z)$ satisfies the hypothesis (14.26) of Lemma 14.3.5. Furthermore,

$$\Delta Q(z) = (\Delta u)_y = 8e^{2u}u_y = 8\rho_{01}^2(z)Q(z).$$

Therefore we may take $p(z) = 8\rho_{01}^2(z)$ in Lemma 14.3.5, and deduce that $\lambda(z) \geq 0$ so that $Q(z) \leq 0$ for all z in the upper half plane and therefore that inequality (14.38) holds. The symmetry relation (14.40) yields $Q(z) \geq 0$ in the lower half plane so that inequality (14.39) holds and the proposition is proved. $\square$

14.3.4 The existence of a lower bound on ρ_{01}

We are now ready to prove the existence of an important lower bound for the hyperbolic density ρ_{01}.

Theorem 14.3.1 *There exists a constant $C > 0$ such that*

$$\rho_{01}(z) \geq \frac{1}{2|z|\,|\log|z|| + 2C|z|}$$

for all z in $\mathbb{C} \setminus \{0, 1\}$ where $C = \frac{1}{2\rho_{01}(-1)}$.

In Exercise 14.2 the value of $\rho_{01}(-1)$ is estimated to be $.114 > .1$. We can thus write the estimate as

$$\rho_{01}(z) \geq \frac{1}{2\left|z\log\frac{1}{|z|}\right| + 10|z|} \tag{14.43}$$

for all z in $\mathbb{C} \setminus \{0, 1\}$.

Proof of Theorem. By Proposition 14.3.1 it is enough to show that, for all $r > 0$,

$$\rho_{01}(-r) \geq \frac{1}{r\left(|2\log r| + \frac{1}{\rho_{01}(-1)}\right)}. \tag{14.44}$$

Let

$$w(\sigma) = u(e^{\sigma + i\pi}) + \sigma, \tag{14.45}$$

where again $u = \log \rho_{01}$ and $\sigma = \log r$. Thus

$$w(\sigma) = \log \rho_{01}(-r) + \log r = \log(r\rho_{01}(-r))$$

and inequality (14.44) is equivalent to

$$e^{-w(\sigma)} \leq 2|\sigma| + e^{-w(0)}. \tag{14.46}$$

Differentiating (14.45) we obtain

$$w'(\sigma) = R(-r) + 1. \tag{14.47}$$

Therefore inequalities (14.36) and (14.37) in the proof of Proposition 14.3.1 imply

$$w'(\sigma) \leq 0 \text{ when } \sigma > 0 \tag{14.48}$$

and

$$w'(\sigma) \geq 0 \text{ when } \sigma < 0.$$

Furthermore, Lemma 14.3.3 implies

$$w'(\sigma) \to 0 \text{ as } \sigma \to \infty. \tag{14.49}$$

From Proposition 14.3.1 we deduce

$$\frac{\partial^2 u}{\partial \theta^2}(-r) \geq 0.$$

Recalling equation (14.31) we get

$$\frac{\partial^2 u}{\partial \sigma^2}(-r) \leq 4e^{2(\sigma + u)}.$$

Equivalently,

$$w'' \leq 4e^{2w}. \tag{14.50}$$

An easy calculation for $\sigma > 0$ shows that (14.48) and (14.50) yield

$$(w'^2 - 4e^{2w})' \geq 0. \tag{14.51}$$

Now integrate inequality (14.51) from $\sigma \geq 0$ to ∞. For the upper limit we invoke Lemma 14.3.4 to deduce that

$$4e^{2w} = 4r^2 \rho_{01}^2(-r) \to 0 \text{ as } \sigma \to \infty \qquad (14.52)$$

and invoke relation (14.49) to deduce that

$$4e^{2w(\sigma)} - w'(\sigma)^2 \geq 0$$

when $\sigma > 0$. Taking square roots and using (14.48) we get

$$0 \geq w'(\sigma) \geq -2e^{w(\sigma)}$$

for $\sigma > 0$. Thus,

$$-e^{-w(\sigma)}w'(\sigma) \leq 2 \text{ when } \sigma > 0.$$

Integrating from 0 to σ yields

$$e^{-w(\sigma)} - e^{-w(0)} \leq 2\sigma,$$

which is just inequality (14.46) when σ is positive. The argument for σ negative follows in the same manner. $\square$

14.4 Properties of the smallest hyperbolic density

In the last section we found a lower bound on the growth of ρ_{01} in a neighborhood of the origin. We can get an upper bound by observing that the punctured disk Δ^* is a subset of $\mathbb{C} \setminus \{0, 1\}$. Thus the inclusion mapping together with formula (7.17) yields

$$\rho_{01}(z) \leq \rho_{\Delta^*}(z) = \frac{1}{2|z| \log \frac{1}{|z|}}, \qquad (14.53)$$

for all z with $0 < |z| < 1$.

To obtain similar estimates near the the other two boundary points, 1 and ∞, we apply the Möbius transformations $z \mapsto 1 - z$ and $z \mapsto \frac{1}{z}$ that preserve $\mathbb{C} \setminus \{0, 1\}$. Before stating the formulas that will be handy later, let us decide where we should apply them. That is, how do we divide $\mathbb{C} \setminus \{0, 1\}$ into three punctured neighborhoods of 0, 1 and ∞? We would like to obtain three estimates, one for each of the boundary points 0, 1 and ∞. The first estimate should apply to the domain Ω_0 consisting of the points that are closer to 0 than to 1 or ∞. Similarly, the second estimate should apply to the domain Ω_1 consisting of the points that are closer to 1 than to 0 or ∞. Finally, the hird estimate should apply to the domain Ω_∞ consisting of the points that are closer to ∞ than to 1 or 0.

One way to obtain these sets is to take the horizontal line $l = \{z = x + iy \mid x = 1/2\}$ separating 0 and 1. Any point z whose real part is less than $1/2$ is closer to 0 than to 1 in the Euclidean metric. Likewise, any point z whose real part is more than $1/2$ is closer to 1 than to 0 in the Euclidean metric. Of course, we cannot use the Euclidean metric to separate the point ∞ from either 0 or 1. We may, however, use the Möbius transformations $f(z) = 1/z$ and $g(z) = (z-1)/z$ that preserve $\mathbb{C} \setminus \{0, 1\}$ and map ∞ to 0 and 1, respectively. Observe that $f(l)$ is the circle $S(1, 1)$ with center at 1 and radius 1 and that $g(l)$ is the unit circle $S(0, 1)$. The three curves l, $S(0, 1)$ and $S(1, 1)$ determine the three domains Ω_0, Ω_1 and Ω_∞.

$$\Omega_0 = \left\{z = x + iy \mid x < \frac{1}{2} \text{ and } 0 < |z| < 1\right\},$$

$$\Omega_1 = \left\{z = x + iy \mid x > \frac{1}{2} \text{ and } 0 < |z - 1| < 1\right\},$$

$$\Omega_\infty = \{z \mid |z - 1| > 1 \text{ and } |z| > 1\},$$

Now we are ready to get the estimates. If z is in Ω_0, then combining formulas (14.43) and (14.53) we obtain

$$\frac{1}{2|z| \log \frac{1}{|z|} + 10|z|} \leq \rho_{01}(z) \leq \rho_{\Delta^*}(z) = \frac{1}{2|z| \log \frac{1}{|z|}}. \tag{14.54}$$

Applying the Möbius transformation $f(z) = 1/z$ to the estimate (14.54) we obtain

$$\frac{1}{2|z| \log |z| + 10|z|} \leq \rho_{01}(z) \leq \frac{1}{2|z| \log |z|} \tag{14.55}$$

for all $z \in \Omega_\infty$. Applying the Möbius transformation $z \mapsto 1 - z$ to the estimate (14.54) we obtain

$$\frac{1}{2|1 - z| \log \frac{1}{|1-z|} + 10|1 - z|} \leq \rho_{01}(z) \leq \frac{1}{2|1 - z| \log \frac{1}{|1-z|}} \tag{14.56}$$

for all $z \in \Omega_1$. The estimate (14.54), together with the inequality

$$\log \frac{1}{|z|} \geq \log 2$$

whenever $|z| < \frac{1}{2}$, has the following corollary.

Corollary 14.4.1 *There is a constant K such that*

$$\frac{1}{K} \le \rho_{01}(z)|z| \log \frac{1}{|z|} \le \frac{1}{2} \le K$$

for all z with $0 < |z| < \frac{1}{2}$.

Note that, since $\log 2 > .69$, the constant K is approximately 16.

Remark. This corollary is just the statement that ρ_{01} is equivalent to ρ_{Δ^*} in the punctured disk $0 < |z| < \frac{1}{2}$.

We end this section with a theorem that compares the values of ρ_{01} at two different points.

Theorem 14.4.1 *Let* $f(z) = |1 - z|\rho_{01}(z)$. *Then there exists a constant D such that*

$$\frac{f(z)}{f(w)} \le D$$

whenever $|w| \le 2|z|$ *and* $|w - 1| \ge \frac{1}{4}$.

Proof. Let K be the constant in Corollary 14.4.1. By the continuity of the Poincaré density proved in Theorem 7.2.2, the function $f(z)$ is nonzero, real-valued, and continuous in $\mathbb{C} \setminus \{0, 1\}$. The formula in Corollary 14.4.1 implies that $f(z)$ extends continuously to the origin with $f(0) = \infty$. Applying the Möbius transformations $z \mapsto 1/z$ and $z \mapsto 1 - z$ to $f(z)$ and using the formula in Corollary 14.4.1 we deduce that, when $|z - 1| \le 1/2$,

$$f(z) = |1 - z|\rho_{01}(1 - z) \le \frac{K}{\log \frac{1}{|1-z|}},$$

and, when $|z| \ge 2$,

$$f(z) = |1 - z|\rho_{01}(1/z) \frac{1}{|z|^2} \le |z - 1| \frac{K}{|z| \log |z|}.$$

Thus, $f(z)$ extends continuously to the whole extended complex plane with $f(\infty) = f(1) = 0$ and we have

$$M = \max \left\{ f(z) \mid |z| \ge \frac{1}{8} \right\} < \infty$$

and

$$N = \min \left\{ f(z) \mid |z| \le \frac{1}{4} \right\} > 0.$$

Therefore, if $|w| \leq 1/4$ and $|z| \geq 1/8$, then

$$\frac{f(z)}{f(w)} \leq \frac{M}{N}.$$

On the other hand, if $|w| \leq 1/4$ and $|z| \leq 1/8$, then Corollary 14.4.1 implies

$$\frac{f(z)}{f(w)} \leq K^2 \frac{|w| \log \frac{1}{|w|}}{|z| \log \frac{1}{|z|}} \frac{|1-z|}{|1-w|} \leq \frac{3K^2}{2} \frac{|w| \log \frac{1}{|w|}}{|z| \log \frac{1}{|z|}}.$$

Since the function $x \mapsto x \log \frac{1}{x}$ is increasing on the interval $(0, e^{-1})$, if $|w| \leq 2|z|$ we have

$$\frac{f(z)}{f(w)} \leq \frac{3K^2}{2} \frac{|2z| \log \frac{1}{|2z|}}{|z| \log \frac{1}{|z|}} \leq 3K^2.$$

This gives the result for $|w| \leq 2|z|$ and $|w| \leq 1/4$.

We break the rest of the proof into two parts. As above assume that $|w| \leq 2|z|$ and now also assume that $|w - 1| \geq 1/4$. We first assume, in addition, that $1/4 \leq |w| \leq 8$. Then the point w is away from $0, 1$ and ∞ and so $f(w) \geq L$ for some $L > 0$. We also have $|z| \geq |w|/2 \geq 1/8$ so that $f(z) \leq M$. Thus the quotient

$$\frac{f(z)}{f(w)} \leq \frac{M}{L},$$

and the theorem follows in this case.

Finally, assume that $|w| \geq 8$ so that $|z| \geq |w|/2 \geq 4$. Corollary 14.4.1 yields

$$\rho_{01}\left(\frac{1}{z}\right) \frac{1}{|z|} \log |z| \leq K$$

and

$$\frac{1}{K} \leq \rho_{01}\left(\frac{1}{w}\right) \frac{1}{|w|} \log |w|.$$

Thus,

$$\rho_{01}(z)|z| \log |z| \leq K$$

and

$$\frac{1}{K} \leq \rho_{01}(w)|w| \log |w|.$$

Therefore we have

$$\frac{f(z)}{f(w)} \leq K^2 \frac{1+|z|}{|w|-1} \frac{|w|\log|w|}{|z|\log|z|} = K^2 \frac{1+\frac{1}{|z|}}{1-\frac{1}{|w|}} \frac{\log|w|}{\log|z|}$$

$$\leq \frac{10K^2}{7} \frac{\log|w|}{\log|z|} \leq \frac{10K^2}{7} \frac{\log|2z|}{\log|z|}$$

$$\leq \frac{10K^2}{7} \left(1 + \frac{\log 2}{\log 4}\right) = \frac{15K^2}{7} \leq 3K^2,$$

which completes the proof. $\square$

14.5 Comparing Poincaré densities

Two metrics ρ_1, ρ_2 defined on a domain are equivalent if there are constants K_1, K_2 such that

$$K_1 \leq \frac{\rho_2}{\rho_1} \leq K_2.$$

It turns out that the pair density $\rho_{ab}(z)$, taken as the supremum over all pairs of distinct points in the complement of Ω, is equivalent to the hyperbolic density on every hyperbolic domain Ω. This is what we shall show in this section.

Since $\Omega \subset \mathbb{C} \setminus \{a, b\}$ we always have

$$\frac{\rho_\Omega}{\rho_{ab}} \geq 1$$

so that we need only find an upper bound.

First we provide an example of a class of domains where it is relatively easy to find a constant for the upper bound.

Theorem 14.5.1 *Suppose the boundary of Ω is connected and unbounded. Then*

$$\rho_\Omega(z) \leq 2C\rho_{ab}(z)$$

for every z in Ω where C is the constant $(2\rho_{01}(-1))^{-1}$.

Proof. Let z be any point in Ω. Pick a point a on the boundary of Ω that is closest to z in the Euclidean metric. Let S_a be the circle with center at a and radius $|z-a|$. By hypothesis, the boundary of Ω contains the points a and ∞ and some set connecting them; thus there must be a point b in S_a such hat b also belongs to the boundary of Ω. Applying the affine transformation $w \mapsto (w-a)/(b-a)$ to Ω, we obtain

$$\rho_{ab}(z) = \frac{1}{|b-a|}\rho_{01}\left(\frac{z-a}{b-a}\right).$$

Now, since $|z-a| = |b-a|$, Theorem 14.3.1 implies

$$\rho_{01}\left(\frac{z-a}{b-a}\right) \geq \frac{|b-a|}{2C|z-a|}.$$

Thus

$$\rho_{ab}(z) \geq \frac{1}{2C|z-a|}.$$

Finally, formula (14.1) implies

$$\rho_\Omega(z) \leq \frac{1}{|z-a|}$$

and therefore

$$\rho_\Omega(z) \leq 2C\rho_{ab}(z). \quad \square$$

Now we show that the hyperbolic density is equivalent to the pair density for all plane domains. Moreover, the constant for the equivalence is universal.

Theorem 14.5.2 *([19],[64]) There exists a constant B such that, for every hyperbolic plane domain Ω, point z in Ω and point a on the boundary of Ω closest to z, there exists a point b on the boundary of Ω such that*

$$\rho_\Omega(z) \leq B\rho_{ab}(z).$$

Proof. Let z be any point in the plane domain Ω and let a be any point on the boundary of Ω closest to z in the Euclidean metric. Let S_a be the circle $|\zeta - a| = |z - a|$. If there is a point b on $\partial\Omega$ such that $b \in S_a$, then the argument of the above theorem shows that

$$\frac{\rho_\Omega(z)}{\rho_{ab}(z)} \leq 2C.$$

Suppose there are no such points. This can happen, for example, if a lies on a bounded component γ of $\partial\Omega$. We will argue as follows. We will find a point b on $\partial\Omega$ and an annulus centered at a, contained in Ω, such that b lies on its boundary and z lies on its core curve. Since we know the hyperbolic density for an annulus, we can use the inclusion map to get the estimate.

Choose a point b on the boundary of Ω such that $\left|\frac{b-a}{z-a}\right|$ is as close to 1 as possible. Equivalently, choose b so that

$$c = \left| \log \left| \frac{b-a}{z-a} \right| \right|$$

is a minimum. By Theorem 14.3.1,

$$\rho_{01}(z) \geq \frac{1}{2 \left| z \log \frac{1}{|z|} \right| + 2C|z|}.$$

Applying the affine transformation $w \mapsto \frac{w-a}{b-a}$ to z we obtain

$$\rho_{ab}(z) \geq \frac{1}{|z-a|(2|\log|\frac{b-a}{z-a}|| + 2C)},$$

and

$$|z-a|\rho_{ab}(z) \geq \frac{1}{2c+2C}. \tag{14.57}$$

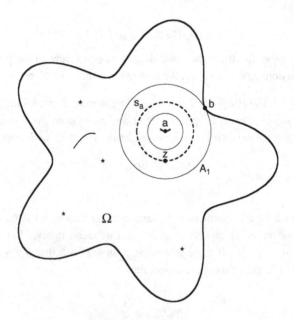

Figure 14.1   The domain Ω with points z, a, b

Suppose first that $|b-a| > |z-a|$, so that b lies outside S_a. Then, by the choice of b, every $w \in \partial\Omega$ outside S_a satisfies

$$|w-a| \geq |b-a|.$$

Moreover, every $w \in \partial\Omega$ inside S_a satisfies

$$\left|\frac{b-a}{z-a}\right| \le \left|\frac{z-a}{w-a}\right|.$$

Therefore none of the points in the annulus

$$A_1 = \left\{ w \mid \frac{|z-a|^2}{|b-a|} < |w-a| < |b-a| \right\}$$

can lie on $\partial\Omega$. Since the point z belongs to both A_1 and Ω, we conclude that $A_1 \subset \Omega$. The inclusion mapping from A_1 to Ω implies that

$$\rho_\Omega(z) \le \rho_{A_1}(z).$$

Note that the point z belongs to the core curve of the annulus A_1. Applying the affine map $w \mapsto \frac{w-a}{b-a}$ to the annulus we obtain

$$\rho_{A_1}(z) = \frac{1}{|b-a|}\rho_{A_x}\left(\frac{z-a}{b-a}\right)$$

where $A_x = \{w \mid x < |w| < 1\}$ and $x = |\frac{z-a}{b-a}|^2$. The inclusion map, together with formula (7.18) for the density of the annulus, yields

$$\rho_\Omega(z) \le \frac{\pi}{4|z-a|\log\left|\frac{b-a}{z-a}\right|} = \frac{\pi}{4|z-a|c}.$$

This, together with formula (14.1), yields

$$|z-a|\rho_\Omega(z) \le \min\left\{1, \frac{\pi}{4c}\right\}. \qquad (14.58)$$

Combining (14.57) and (14.58) we obtain

$$\frac{\rho_\Omega(z)}{\rho_{ab}(z)} \le \min\left\{2c+2C, \frac{\pi(2c+2C)}{4c}\right\}.$$

The function $f(c) = 2c+2C$ is increasing in c and the function $g(c) = \frac{\pi(2c+2C)}{4c}$ is decreasing in c and at the point $c = \pi/4$, $f(\pi/4) = g(\pi/4)$. Thus,

$$\frac{\rho_\Omega(z)}{\rho_{ab}(z)} \le \frac{\pi}{2} + 2C = B.$$

Approximating $2C$ by 10 gives the estimate $B < 11.58$.

Suppose now that $|b-a| < |z-a|$, so that b lies inside S_a. By the choice of b, every $w \in \partial\Omega$ inside S_a satisfies

$$|w-a| \le |b-a|$$

and every $w \in \partial\Omega$ outside S_a satisfies

$$\left| \frac{w-a}{z-a} \right| \geq \left| \frac{z-a}{b-a} \right| > 1.$$

Therefore there are no points of $\partial\Omega$ in the annulus

$$A_2 = \left\{ w \mid |b-a| < |w-a| < \frac{|z-a|^2}{|b-a|} \right\}.$$

Since the point z belongs to both A_2 and Ω, we conclude that $A_2 \subset \Omega$. Thus the inclusion mapping from A_2 to Ω yields

$$\rho_\Omega(z) \leq \rho_{A_2}(z).$$

The point z belongs to the core curve of the annulus A_2, and that again simplifies the formula for the hyperbolic density. Applying the affine map $w \mapsto \frac{(w-a)(b-a)}{(z-a)^2}$ we obtain

$$\rho_{A_2}(z) = \frac{|b-a|}{|z-a|^2} \rho_{A_y} \left(\frac{b-a}{z-a} \right)$$

where $y = \left| \frac{b-a}{z-a} \right|^2$. Formula (7.18) now yields

$$\rho_\Omega(z) \leq \frac{\pi}{4|z-a| \log \left| \frac{z-a}{b-a} \right|} = \frac{\pi}{4|z-a|c}.$$

Therefore the same arguments as above give us the estimate

$$\frac{\rho_\Omega(z)}{\rho_{ab}(z)} \leq B = 2C + \frac{\pi}{2}. \quad \square$$

Exercise 14.4 Evaluate one possible value for the constant D of Theorem 14.4.1.

15

Uniformly perfect domains

In Chapter 10 we found that the easiest estimate from above on the hyperbolic density was obtained by using the inclusion map from the largest disk centered at the point in question. We saw that, if t is this point and if $d(t)$ is the Euclidean distance from t to the boundary of Ω, then

$$\rho_\Omega(t) \le \frac{1}{d(t)}.$$

We call the density function $\frac{1}{d(t)}$ the *quasi-hyperbolic density*. Note that it, like the hyperbolic density, is unbounded in the domain. It is often possible to get further estimates for domains whose hyperbolic density is equivalent to the $\frac{1}{d(t)}$ density in the sense that there is a constant $Q = Q(\Omega)$, depending only on the domain, such that

$$\frac{1}{Qd(t)} \le \rho_\Omega(z) \le \frac{1}{d(t)}.$$

These domains have a special name.

Definition 15.1 *A hyperbolic domain Ω is* uniformly perfect *if the hyperbolic density is equivalent to the quasi-hyperbolic density.*

One simple example of a uniformly perfect domain is the unit disk Δ where it is easy to check that

$$\frac{1}{2d} \le \rho(t) \le \frac{1}{d}.$$

The punctured disk, as we will see below, is an example of a domain that is not uniformly perfect.

In Chapter 7 we derived formula (7.17) for the hyperbolic density of the punctured unit disk Δ^*. Recall that

$$\rho_{\Delta^*}(t) = \frac{1}{2|t|\log\frac{1}{|t|}}.$$

Observe that, if $|t| \geq 1/2$, then the distance $d(t)$ from the point t to the boundary of Δ^* is equal to $1 - |t|$ and

$$\rho_{\Delta^*}(t) < \frac{1}{1-|t|}.$$

Now observe that, if $|t| < 1/2$, then the distance $d(t)$ from the point t to the boundary of Δ^* is equal to $|t|$. Therefore, although $\rho_{\Delta^*}(t)$ approaches infinity as t approaches the boundary component $\{0\}$ of Δ^*, $|t|\rho_{\Delta^*}(t)$ approaches zero. Thus, there is no universal constant $C > 0$ such that

$$\rho_{\Delta^*}(t) \geq \frac{C}{d(t)}$$

for all t in Δ^*.

In addition to the discussion in this chapter, the reader is referred to the literature on uniformly perfect domains, for example, [55], [7], [52], [21].

15.1 Simple examples

We saw above that the punctured disk is not uniformly perfect. More generally any punctured domain is not uniformly perfect.

Proposition 15.1.1 *If a hyperbolic domain Ω has a puncture, then Ω is not uniformly perfect.*

Proof. Suppose that Ω is a hyperbolic domain and that c is a puncture of Ω. Then there exists a punctured disk D^* with center at c such that $D^* \subset \Omega$. Therefore,

$$\rho_\Omega(z) \leq \rho_{D^*}(z)$$

for every z in D^*. From formula (7.17) we can easily derive

$$\rho_{D^*}(z) = \frac{1}{2|z-c|\log\frac{r}{|z-c|}}$$

where r is the Euclidean radius of D^*. Thus, as z approaches c,

$$\rho_{D^*}(z)d(z) = \rho_{D^*}(z)|z-c| = \frac{1}{2\log\frac{r}{|z-c|}} \to 0$$

and Ω is not uniformly perfect. □

While Proposition 15.1.1 provides many examples of non-uniformly-perfect domains, the next proposition quickly provides many uniformly perfect examples.

Proposition 15.1.2 *If the boundary of Ω, together with the point at infinity, is a connected set in the Riemann sphere, then Ω is uniformly perfect.*

Proof. Let z be any point in Ω. Pick a point a on the boundary of Ω that is closest to z; that is, such that $d(z) = |z - a|$. By the proof of Theorem 14.5.1 there exists a point b on the boundary of Ω such that $|z - a| = |b - a|$. Since Ω is a subset of $\mathbb{C} \setminus \{a, b\}$ we have

$$\rho_\Omega(z) \geq \rho_{ab}(z).$$

On the one hand, applying the Möbius transformation $w \mapsto \frac{w-a}{b-a}$ we obtain

$$\rho_\Omega(z) \geq \frac{1}{|b - a|} \rho_{01}\left(\frac{z - a}{b - a}\right),$$

and therefore

$$\rho_\Omega(z)d(z) \geq \rho_{01}\left(\frac{z - a}{b - a}\right). \qquad (15.1)$$

On the other hand we have

$$|z - a| = |b - a|.$$

Proposition 14.3.1 thus implies that $\rho_\Omega(z)d(z) \geq \rho_{01}(-1)$. □

15.2 Uniformly perfect domains and cross ratios

Let Ω be any hyperbolic plane domain and let a and b be any two distinct points in the complement of Ω. Given z in Ω, we know that

$$\rho_\Omega(z) \geq \rho_{ab}(z).$$

The points a and b together with the points z and ∞ form a quadruple of distinct points and we may look at the relationship of their cross ratio with $\rho_\Omega(z)$.

For cross ratios, we make the definition

Definition 15.2 *The cross ratio*

$$cr(z, a, b, \infty) = \frac{z - a}{b - a}$$

is ϵ-bounded *if*

$$\epsilon < |cr(z, a, b, \infty)| < \frac{1}{\epsilon} \qquad (15.2)$$

and

$$\epsilon < |1 - cr(z, a, b, \infty)|. \qquad (15.3)$$

It would be nice if we were to choose one of the points, say the point a, to be the closest point to z, that is

$$|z - a| \le |z - c| \text{ for all } c \in \mathbb{C} \setminus \Omega. \qquad (15.4)$$

It would be even nicer if we could choose the other point, b, so that the cross ratio $cr(z, a, b, \infty)$ is uniformly bounded. We give a name to domains Ω for which we can pick such a nice pair of points for every point z in Ω.

Definition 15.3 *A hyperbolic plane domain Ω has the* bounded cross ratio *property if there exists $\epsilon > 0$ such that, for every z in Ω and for every point a on the boundary of Ω closest to z, there exists a point b on the boundary of Ω such that $cr(z, a, b, \infty)$ is ϵ-bounded.*

Theorem 15.2.1 *A hyperbolic plane domain Ω is uniformly perfect if and only if Ω has the bounded cross ratio property.*

Proof. As in Theorem 14.4.1, let

$$f(z) = |1 - z|\rho_{01}(z).$$

As in the proof of Theorem 14.4.1, the function $f(z)$ is nonzero, real valued, and continuous in $\mathbb{C} \setminus \{0, 1\}$, and it extends continuously to the whole extended complex plane with $f(0) = \infty$ and $f(\infty) = f(1) = 0$. The function

$$g(z) = |z|\rho_{01}(z) = f(1 - z)$$

is therefore also nonzero, real valued, and continuous in $\mathbb{C} \setminus \{0, 1\}$. It extends continuously to the points $z = 0$, $z = \infty$ and $z = 1$ by setting $g(0) = g(\infty) = 0$ and $g(1) = \infty$.

Suppose first that Ω is uniformly perfect and let z be a point in Ω. Pick a point a on the boundary of Ω closest to the point z. Then by Theorem 14.5.2 there exist a constant C and a point b on the boundary of Ω such that $\rho_{\Omega}(z) \le C\rho_{ab}(z)$. Applying the Möbius transformation $w \mapsto (w - a)/(b - a)$ we obtain

$$\rho_{\Omega}(z) \le \frac{C}{|b - a|}\rho_{01}\left(\frac{z - a}{b - a}\right).$$

Since Ω is uniformly perfect, there exists a constant $Q(\Omega) = \frac{1}{Q(\Omega)} > 0$ such that $k(\Omega) \leq \rho_\Omega(z)|z-a|$. Thus

$$\left|\frac{z-a}{b-a}\right| \rho_{01}\left(\frac{z-a}{b-a}\right) \geq \frac{Q(\Omega)}{C}. \tag{15.5}$$

In other words,

$$g\left(\frac{z-a}{b-a}\right) \geq \frac{Q(\Omega)}{C}.$$

Since $g(0) = g(\infty) = 0$, this implies that $cr(z, a, b, \infty) = (z-a)/(b-a)$ is bounded away from 0 and ∞. Finally, because of condition (15.4), we see that

$$|1 - cr(z, a, b, \infty)| = \left|\frac{z-b}{b-a}\right| \geq \left|\frac{z-a}{b-a}\right|$$

is also bounded away from zero. Therefore Ω has the bounded cross ratio property.

Suppose next that Ω has the bounded cross ratio property. Let z be a point in Ω and pick a closest point a on the boundary of Ω. We want to show that $|z-a|\rho_\Omega(z)$ is uniformly bounded below. Since Ω has the bounded cross ratio property, there exist an $\epsilon > 0$ and a point b on the boundary of Ω such that $cr(z, a, b, \infty)$ is ϵ-bounded. Therefore, we have

$$\rho_\Omega(z)|z-a| \geq \rho_{ab}(z)|z-a|$$
$$= \left|\frac{z-a}{b-a}\right| \rho_{01}\left(\frac{z-a}{b-a}\right) \geq \epsilon\rho_{01}\left(\frac{z-a}{b-a}\right)$$

and, because the cross ratio $\frac{z-a}{b-a}$ is ϵ-bounded, this is uniformly bounded away from zero and Ω is uniformly perfect. $\square$

15.3 Uniformly perfect domains and separating annuli

Recall from subsection 7.6.2 that the modulus of a round annulus $R = \{z \mid a \leq |z-c| \leq b\}$ is defined as $\mathrm{mod}(R) = \frac{1}{2\pi}\log\frac{b}{a}$. Moreover, given any doubly connected domain D, it is conformally equivalent to some round annulus $R(D)$. The modulus of D is defined as $\mathrm{mod}(D) = \mathrm{mod}(R(D))$.

The modulus thus measures the shape of the annulus – that is, how fat or thin it is. In the computations that follow it is easier to work with the quantity $b/a = \exp 2\pi \,\mathrm{mod}\,(D)$ which also measures the shape.

Definition 15.4 *Let Ω be a hyperbolic plane domain. An annulus $A \subset \Omega$ is called a* separating annulus *if A separates boundary components of Ω.*

Define

$$A(\Omega) = \sup_{D} \exp 2\pi \bmod (D)$$

where the supremum is taken over all separating annuli D in Ω. Similarly, define

$$RA(\Omega) = \sup_{R} \exp 2\pi \bmod (R)$$

where the supremum is taken over all separating round annuli R in Ω. Obviously $RA(\Omega) \le A(\Omega)$.

These two quantities are equivalent; that is,

Theorem 15.3.1 $A(\Omega)$ *is bounded if and only if* $RA(\Omega)$ *is.*

This theorem is useful because it is often possible to check for bounds on round separating annuli but not on arbitrary separating annuli. The proof is based on the following lemmas.

Lemma 15.3.1 *Let* $A_n = \{z | 1/R_n < |z| < R_n\}$ *and assume* $R_n \to \infty$ *as n goes to ∞. Any sequence of univalent maps* $f_n : A_n \to \mathbb{C} \setminus \{0\}$ *such that* $f_n(A_n)$ *separates* 0 *and* ∞ *and* $f_n(1) = 1$ *is a normal family and any convergent subsequence* f_{n_j} *converges to either the identity or the reciprocal map* $f(z) = 1/z$ *as n_j goes to ∞.*

Proof. The f_n are univalent on A_n and so the f_n restricted to $A_n \setminus \{1\}$ are also. The restricted maps omit $0, 1, \infty$ and so form a normal family by Montel's theorem. Using the diagonal method, we can find a limit function f defined on $\mathbb{C} \setminus \{0, 1\}$. If γ is the circle $|z| = 1/2$ it is contained in all the punctured annuli for large n. The images $f_n(\gamma)$ separate 0 and ∞ so any limit function f of a convergent subsequence also separates them and cannot be constant. By Hurwitz' theorem, we conclude that f must be a univalent map of $\mathbb{C} \setminus \{0, 1\}$ to itself. By Theorem 3.1.3, f extends to a Möbius map and, since $f_n(1) = 1$, f has to fix 1. It either fixes 0 and ∞ or interchanges them so is either the identity or the reciprocal map. $\square$

Lemma 15.3.2 *Let* A_n *be a sequence of annuli whose moduli converge to infinity. Then each* A_n *contains a round annulus* R_n *and the moduli of the* R_n *also converge to infinity.*

Proof. By Theorem 7.6.3 there is a univalent map f_n from a round annulus $R_n = \{z | 1/r_n < |z| < r_n\}$ onto A_n such that $r_n \to \infty$ as $n \to \infty$. With no loss of generality, we may assume that the A_n separate 0 from ∞ and that

$f_n(1) = 1$. By Lemma 15.3.1, the f_n form a normal family and any convergent subsequence has either the identity or the reciprocal map as a limit.

Fix $N > 1$. We claim the annulus $\mathcal{A} = \{z \mid 1/N < |z| < N\}$ must be contained in $f_n(R_n)$ for all sufficiently large n. By Lemma 15.3.1, any convergent subsequence of the f_n is either close to the identity or the reciprocal map. For large n, therefore, the images of the circles $|z| = 1/2N$ and $|z| = 2N$ are curves close to these circles and so bound a domain containing $\mathcal{A}$, proving the claim. $\square$

Proof of Theorem 15.3.1. It follows from the definitions that if $A(\Omega)$ is bounded then $RA(\Omega)$ is bounded. Suppose $A(\Omega)$ is unbounded. Then there is a sequence of separating annuli $A_n \subset \Omega$ whose moduli go to infinity. By Lemma 15.3.2, we can find a sequence of round separating annuli $R_n \subset A_n \subset \Omega$ whose moduli also go to infinity so that $RA(\Omega)$ is unbounded. $\square$

The next theorem says that uniformly perfect domains are those that have a bound on the moduli of separating round annuli or a bound on the moduli of arbitrary separating annuli.

Theorem 15.3.2 *The following three properties are equivalent for any hyperbolic plane domain Ω:*

$$\Omega \text{ is uniformly perfect;} \tag{15.6}$$

$$A(\Omega) < \infty; \tag{15.7}$$

$$RA(\Omega) < \infty. \tag{15.8}$$

Proof. Theorem 15.3.1 says that conditions (15.7) and (15.8) are equivalent. It therefore suffices to show that (15.6) and (15.8) are equivalent.

Suppose first that $RA(\Omega) < \infty$. Let z be any point in Ω and let a be a point on the boundary of Ω closest to z. Let X be the round annulus with modulus $\frac{1}{2\pi}\log\left(RA(\Omega) + 1\right)$ defined by

$$X = \left\{ w \mid \frac{1}{\sqrt{RA(\Omega)+1}} |z - a| \le |w - a| \le \sqrt{RA(\Omega)+1}\,|z - a| \right\}.$$

Then X is not separating for Ω but it does, however, separate the point a from the point at infinity, and, since $RA(\Omega) < \infty$, it must intersect the boundary of Ω. Let b be a point in that intersection. Then, since b belongs to X, the cross ratio $cr = cr(z, a, b, \infty) = \frac{z-a}{b-a}$ satisfies

$$\frac{1}{\sqrt{RA(\Omega)+1}} \leq |cr| \leq \sqrt{RA(\Omega)+1}.$$

Furthermore,

$$|1-cr| = \left|\frac{z-b}{a-b}\right| \geq |cr| \geq \frac{1}{\sqrt{RA(\Omega)+1}}.$$

Therefore, Ω has the bounded cross ratio property and is uniformly perfect by Theorem 15.2.1.

Suppose next that Ω is uniformly perfect. Let X be a round annulus inside Ω which separates the boundary of Ω. Let p be the center of X. By expanding X in either direction, if necessary, we get the biggest possible round annulus Y containing X that is still in Ω. That is, there are a minimal c and a maximal d, such that

$$X \subset Y = \{w \mid c \leq |w-p| \leq d\} \subset \Omega.$$

There are thus points a and x on the boundary of Ω such that $|a-p| = c$ and $|x-p| = d$.

Take the point z in Y such that z, a and p are colinear and $|z-p| = \sqrt{cd}$. Then $|z-a|$ is the Euclidean distance from z to the boundary of Y. This implies that $|z-a|$ is also the Euclidean distance from z to the boundary of Ω. Since Ω is uniformly perfect, Theorem 15.2.1 says there is a point b on the boundary of Ω such that $cr(z, a, b, \infty)$ is ϵ-bounded. Since b is not in Ω, it cannot be in Y either. Thus, $|b-p| \leq c$ or $|b-p| \geq d$.

If $|b-p| \leq c$, then we have

$$\frac{1}{\epsilon} \geq |cr| \geq \frac{|z-a|}{|b-p|+|a-p|} \geq \frac{\sqrt{cd}-c}{2c} = \frac{1}{2}\sqrt{\frac{d}{c}} - \frac{1}{2}.$$

Thus $\mathrm{mod}(Y)$ is at most $\frac{1}{\pi}\log(\frac{2}{\epsilon}+1)$. Since X is a round annulus with the same center as Y and contained in Y, $\mathrm{mod}(X) \leq \mathrm{mod}(Y)$.

If $|b-p| \geq d$, then

$$\epsilon \leq |cr| \leq \frac{\sqrt{cd}}{|a-b|} \leq \frac{\sqrt{cd}}{d-c}.$$

A simple calculation shows that

$$\frac{d}{c} \leq \frac{(1+\sqrt{1+4\epsilon^2})^2}{4\epsilon^2}$$

so that $\mathrm{mod}(Y)$ and hence $\mathrm{mod}(X)$ are again bounded. Therefore in both cases there is a bound on $\mathrm{mod}(X)$. Since X was arbitrary,

$$RA(\Omega) \le \max\left\{ \left(\frac{2}{\epsilon}+1\right)^2, \frac{(1+\sqrt{1+4\epsilon^2})^2}{4\epsilon^2} \right\}.$$

Therefore,

$$RA(\Omega) \le \left(\frac{2}{\epsilon}+1\right)^2. \quad \square$$

15.4 Uniformly thick domains

In this section we will look at another related condition for domains, the boundedness of the injectivity radius.

Definition 15.5 *For any z in* Ω, *set*

$$i(\Omega, z) = \sup_{a}\{\zeta \in \Omega \,|\, \rho\Omega(z,\zeta) < a\} \ \ is \ simply \ connected \ in \ \Omega.$$

Then

$$i(\Omega) = \inf_z i(\Omega, z)$$

is the injectivity radius of the domain Ω.

For example, if Ω is the unit disk Δ, then any hyperbolic disk inside Δ is a Euclidean disk, and so is simply connected. Thus, if Ω is simply connected, then

$$i(\Omega) = \infty. \tag{15.9}$$

As another example, let Ω be the punctured disk Δ^*. For any z in Ω, let l_z be a Euclidean circle with center at 0 and radius $|z|$. The hyperbolic length of l_z,

$$\rho_\Omega(l_z) = \int_{l_z} \rho_\Omega(w)|dw| = \int_{l_z} \frac{1}{2|w|\log\frac{1}{|w|}}|dw|$$

$$= \int_0^{2\pi} \frac{d\theta}{2\log\frac{1}{|z|}} = \frac{\pi}{\log\frac{1}{|z|}},$$

converges to zero as z converges to zero. The hyperbolic disk $D(z, R)$ with center at z and radius $R > \frac{\pi}{\log\frac{1}{|z|}}$ in Ω contains the whole curve l_z. Therefore $D(z, R)$ is not simply connected,

$$i(\Omega, z) \le \frac{\pi}{\log\frac{1}{|z|}},$$

and $i(\Omega) = 0$.

The injectivity radius $i(\Omega, z)$ measures how "thick" the domain Ω is in the neighborhood of the point z. This leads us to define

Definition 15.6 *A domain* Ω *is* uniformly thick *if* $i(\Omega) > 0$.

As we just saw, simply connected domains are uniformly thick, but punctured simply connected domains are not uniformly thick. The reason why punctured disks are not uniformly thick is that they contain non-homotopically-trivial closed curves of arbitrary small hyperbolic length (recall Theorem 7.2.5).

Let z be a point in Ω and let π be a universal covering map from the unit disk Δ onto Ω such that $\pi(0) = z$. Let HD_a be the hyperbolic disk in Δ with center at zero and radius a. Theorem 7.1.2 implies that the image of HD_a is the hyperbolic disk $D(z, a)$ in Ω with center at z and radius a.

Let π_D be the restriction of π to HD_a. If π_D is one to one, then $D(z, a)$ is simply connected because it is a conformal image of a simply connected set and $i(\Omega, z) \geq a$. If, however, π_D is not one to one, then there are two distinct points w_1 and w_2 in HD_a such that $\pi(w_1) = \pi(w_2)$. Let γ be a geodesic in Δ that connects w_1 and w_2. Then, by Exercise 2.18, $\gamma \subset HD_a$, and so $\pi(\gamma)$ is a closed curve in $D(z, a)$. Since π is a universal covering map, $\pi(\gamma)$ is not trivial, and so $D(z, a)$ cannot be simply connected. We conclude, therefore, that a domain Ω is uniformly thick if there exists a positive number a such that the universal covering map π is one to one on every hyperbolic disk in Δ with radius equal to a. Moreover, the largest such a is precisely $i(\Omega)$.

To explore further the connection of uniformly thick domains with non-trivial closed curves, let γ be a non-trivial closed curve in Ω and let $l(\gamma)$ denote its hyperbolic length. Set

$$l(\Omega) = \inf_{\gamma} l(\gamma).$$

If there are no non-trivial closed curves we let $l(\Omega) = \infty$. This only happens when Ω is simply connected, which by (15.9) also means that $i(\Omega) = \infty$. This generalizes further:

Theorem 15.4.1 *For any plane domain* Ω,

$$l(\Omega) = 2i(\Omega).$$

As an immediate corollary we have

Corollary 15.4.1 *A hyperbolic domain* Ω *is* uniformly thick *if and only if*

$$l(\Omega) > 0.$$

Proof. If $k > l(\Omega)$, by the definition of $l(\Omega)$, there exists a non-trivial closed curve γ in Ω whose length is less than k. Take any point z_0 on γ and, as usual, denote by $D(z_0, k/2)$ the hyperbolic disk in Ω with center at z_0 and radius $k/2$. If there is a point w outside $D(z_0, k/2)$ but on γ, then there will be an arc of γ from z_0 to w with length at least $k/2$ and another arc of γ from w back to z_0 with length at least $k/2$. Since this is impossible, $\gamma \subset D(z_0, k/2)$, and, because γ is non-trivial, $D(z_0, k/2)$ cannot be simply connected. It follows that $i(\Omega) \leq k/2$.

For the proof in the other direction, let $k > i(\Omega)$. Then there exists a point z in Ω such that $D(z, k)$ is not simply connected. Let π be a universal covering map from the unit disk Δ onto Ω such that $\pi(0) = z$. Then, the restriction of π to the hyperbolic disk HD_k is not one to one and there are distinct points z_1 and z_2 in HD_k such that $\pi(z_1) = \pi(z_2)$. If γ is a geodesic joining z_1 to z_2, then the curve $\pi(\gamma)$ is a closed non-trivial curve in $D(z, k)$. Furthermore

$$\rho_\Omega(\pi(\gamma)) = \rho(\gamma) = \rho(z_1, z_2) \leq \rho(z_1, 0) + \rho(0, z_2) < 2k$$

and $l(\Omega) \leq 2k$. $\square$

The next theorem shows that uniformly perfect and uniformly thick are equivalent conditions.

Theorem 15.4.2 *A hyperbolic plane domain Ω is uniformly perfect if and only if it is uniformly thick.*

Proof. First we show that if Ω is uniformly thick then it is uniformly perfect. Let z be a point in Ω. Let π be a universal holomorphic covering map from the unit disk Δ onto Ω, with $\pi(0) = z$. Then

$$\rho_\Omega(z) = \frac{1}{|\pi'(0)|}. \tag{15.10}$$

Since Ω is uniformly thick, π is one-to-one on every hyperbolic disk in Δ of radius $i(\Omega)$. Let $HD = HD_{i(\Omega)}$; then this is also a Euclidean disk with center at 0 and Euclidean radius

$$r = \frac{e^{2i(\Omega)} - 1}{e^{2i(\Omega)} + 1}. \tag{15.11}$$

The function

$$g(w) = \frac{\pi(rw) - z}{r\pi'(0)}$$

is therefore a univalent function defined on Δ and we can apply the Koebe $\frac{1}{4}$-theorem (Theorem 3.7.1) to conclude that the image of Δ under g contains the Euclidean disk with center at 0 and radius $1/4$. Hence the image of HD

under π contains the Euclidean disk with center at z and radius $r|\pi'(0)|/4$. Now equations (15.10) and (15.11) imply that the Euclidean disk with center at z and radius

$$\frac{e^{2i(\Omega)} - 1}{4(e^{2i(\Omega)} + 1)\rho_\Omega(z)}$$

is contained inside Ω. Thus, the Euclidean distance $d(z)$ from z to the boundary of Ω satisfies

$$d(z) \geq \frac{e^{2i(\Omega)} - 1}{4(e^{2i(\Omega)} + 1)\rho_\Omega(z)}.$$

Therefore,

$$d(z)\rho_\Omega(z) \geq \frac{e^{2i(\Omega)} - 1}{4(e^{2i(\Omega)} + 1)}.$$

Since Ω is uniformly thick, $i(\Omega) > 0$, and thus Ω is uniformly perfect. Furthermore,

$$Q(\Omega) \leq 4\frac{e^{2i(\Omega)} + 1}{e^{2i(\Omega)} - 1}.$$

Finally we show that, if Ω is uniformly perfect, then $l(\Omega) > 0$, and therefore by Corollary 15.4.1 that Ω is uniformly thick. Choose $\epsilon > 0$ and assume that Ω is uniformly perfect. Pick a non-trivial closed curve γ on Ω. By Theorem 7.2.5, in the free homotopy class of γ either there is a geodesic or there are arbitrarily short curves. The latter cannot happen, though, since $l(\Omega) > 0$. We may assume, therefore, without loss of generality, that γ is geodesic and that $l_\gamma = l(\gamma) \leq l(\Omega) + \epsilon$. By the collar lemma, Lemma 7.7.1, there is a collar $C(\gamma)$ about γ of width

$$w_\gamma = \operatorname{arc\,sinh}\left(\frac{1}{\sinh(l_\gamma)}\right) \geq \operatorname{arc\,sinh}\left(\frac{1}{\sinh((l(\Omega) + \epsilon))}\right) \tag{15.12}$$

and by formula (7.23) the modulus of the collar is $log(l_r/\theta)$ where $\theta = \operatorname{arctanh}(\sinh w_\gamma)$ and is a monotone function of the length l_γ.

By Theorem 15.3.2 the supremum of the moduli of all separating annuli, $A(\Omega)$, is bounded since Ω is uniformly perfect. The collar $C(\gamma)$ is a separating annulus so its modulus (l_r/θ) is bounded. By inequality (15.12), $l(\Omega)$ is bounded below. $\square$

Exercise 15.1 Let Ω be an arbitrary hyperbolic domain. Consider the quantities $Q(\Omega)$ for measuring how uniformly perfect it is, $\epsilon(\Omega)$ for measuring he boundedness of the cross ratio, $A(\Omega)$ for measuring the shapes of its eparating annuli and $i(\Omega)$ for measuring how uniformly thick it is. Compare hese quantities.

Exercise 15.2 Let Ω be the annulus $\{z \mid 1 < |z| < 4\}$. Is Ω uniformly perfect? Evaluate or find estimates on the constants $Q(\Omega)$, $\epsilon(\Omega)$, $A(\Omega)$, $RA(\Omega)$, $i(\Omega)$ and $l(\Omega)$.

Exercise 15.3 Let Ω be Δ minus the standard middle thirds Cantor set. That s, we take the interval $[0, 1]$ and delete the middle third $(1/3, 2/3)$. From he two remaining intervals we again delete the middle thirds etc. The set hat remains is closed and its complement in Δ is Ω. Is Ω uniformly perfect? Evaluate or find estimates on the constants $Q(\Omega)$, $\epsilon(\Omega)$, $A(\Omega)$, $RA(\Omega)$, $i(\Omega)$ and $l(\Omega)$.

Exercise 15.4 Let $\Omega = \{z \mid \frac{1}{4} < |z| < 4 \text{ and } |z - 1| > \frac{1}{4}\}$. Is Ω uniformly perfect? Evaluate or find estimates on the constants $Q(\Omega)$, $\epsilon(\Omega)$, $A(\Omega)$, $RA(\Omega)$, $i(\Omega)$ and $l(\Omega)$.

16

Appendix: a brief survey of elliptic functions

In this appendix, we sketch the results from elliptic function theory that we used in Chapter 14. The reader is referred to standard texts, for example [15], [68] or [57], for detailed discussions.

As motivation, we begin with the cyclic elementary group G_1 defined in Section 5.1 which is a group with elements of the form $z \mapsto z + n$. A holomorphic function $f(z)$ is invariant with respect to the group if it satisfies $f(z+n) = f(z)$, for all $n \in \mathbb{Z}$. An example is the exponential, $E(z) = e^{\pm 2\pi i z}$. It is the covering map from the plane $\mathbb{C}$ to the punctured plane $\mathbb{C} \setminus \{0\}$ and a fundamental domain is the strip $\{z \mid 0 < \Re z < 1\}$. Any holomorphic function or meromorphic function $g(z)$ whose singularities in $\mathbb{C}$ are at most poles, and periodic with respect to G_1, defines a function h on the punctured plane by setting $g(z) = h \circ E(z)$.

We turn now to holomorphic functions periodic under the elementary group $G_{a,b}$ defined in Section 5.1. Recall that it is a group with elements of the form $z \mapsto z + ma + nb$ for some $a, b \in \mathbb{C}, \Im(\frac{b}{a}) > 0$. One fundamental domain for this group is the parallelogram

$$P_{a,b} = \{z \mid z = sa + tb, 0 < s < 1, 0 < t < 1\}.$$

Meromorphic functions invariant under $G_{a,b}$ are the elliptic functions mentioned in Section 6.1; they are also called doubly periodic functions. As for the simply periodic functions above, they define functions on the quotient torus $\mathbb{C}/G_{a,b}$.

16.0.1 Basic properties of elliptic functions

First note that the fundamental domain $P_{a,b}$ for the group $G_{a,b}$ is bounded and hence, by Liouville's theorem, any holomorphic function invariant with respect to the group must be constant.

Any non-constant locally $G_{a,b}$-invariant holomorphic function therefore must have poles and so is meromorphic. Integrating an elliptic function around the boundary of the fundamental domain, it follows by standard arguments that the number of times any value is taken in this domain, counted with multiplicity, is the same.

Weierstrass constructed a function that is easily seen to be elliptic. It is called the *Weierstrass $\wp$-function* and is denoted by the special symbol $\wp$. Given $a, b \in \mathbb{C}$ with $\Im(\frac{b}{a}) > 0$, set $\omega_{n,m} = \{na + mb | n, m \in \mathbb{Z}\}$. Define

$$\wp(z, a, b) = \frac{1}{z^2} + \sum_{(m,n) \neq (0,0)} \left(\frac{1}{(z - \omega_{m,n})^2} - \frac{1}{\omega_{m,n}^2} \right).$$

The poles of this function are clearly at the points $\omega_{m,n}$; this set of points is called the *period lattice*.

To see that the series is analytic away from its poles, note that, for large (m, n), the terms are of order $O(|\omega_{m,n}|^{-3})$ and so the series converges uniformly and absolutely. Since the sum is independent of the order of summation, it has the desired periods. We can choose the fundamental domain so that, for ϵ with small absolute value, $O\epsilon P_{a,b}' = P_{a,b} + \epsilon$. The pole in 1 is at the origin and has order 2 by construction.

Replacing z by $-z$ also does not change the value so $\wp(z; a, b)$ is an even function.

We next look for the critical points. The derivative of $\wp(z, a, b)$ can be computed by term by term differentiation:

$$\wp'(z; a, b) = -2 \sum_{(m,n)} \frac{1}{(z - \omega_{m,n})^3}.$$

This function is also holomorphic away from its poles and has the same periods as $\wp(z; a, b)$. Since both a and b are periods, omitting the periods as arguments, and noting that the difference between two half periods is a full period, we have

$$\wp'\left(-\frac{a}{2}\right) = \wp'\left(\frac{a}{2}\right), \quad \wp'\left(-\frac{b}{2}\right) = \wp'\left(\frac{b}{2}\right) \text{ and } \wp'\left(-\frac{a+b}{2}\right) = \wp'\left(\frac{a+b}{2}\right).$$

The derivative of an even function is an odd function so that

$$\wp'\left(-\frac{a}{2}\right) = -\wp'\left(\frac{a}{2}\right), \quad \wp'\left(-\frac{b}{2}\right) = -\wp'\left(\frac{b}{2}\right) \text{ and } \wp'\left(-\frac{a+b}{2}\right) = -\wp'\left(\frac{a+b}{2}\right).$$

It follows that

$$\wp'\left(\frac{a}{2}\right) = \wp'\left(\frac{b}{2}\right) = \wp'\left(\frac{a+b}{2}\right) = 0.$$

Since $\wp'$ has a single pole of order 3 in a fundamental domain, these are the only zeros.

The critical values of $\wp(z)$ are

$$\wp\left(\frac{a}{2}\right) = e_1, \quad \wp\left(\frac{b}{2}\right) = e_2, \quad \wp\left(\frac{a+b}{2}\right) = e_3.$$

The function $\wp(z) - e_1$ has one double pole in the fundamental domain so it can have only two zeros. It has a double zero at $\frac{a}{2}$ so it cannot have any others, and, in particular, it is not zero at either of the other half periods. The same argument for e_2 and e_3 shows that the three critical values are distinct.

The $\wp$-function is therefore a regular covering from the complex plane onto a two sheeted cover of the Riemann sphere, except at the critical points. The critical values and infinity are branch points of order 2. To make the image $\wp(\mathbb{C})$ into a Riemann surface in the classical way, introduce slits between e_1 and infinity and between e_2 and e_3 on each of the sheets and label the two edges of the slit with a plus or minus. Then, sew the plus edge of the slits on the upper sheet to the minus edge of the corresponding slit on the lower sheet and vice versa. The result is topologically a torus. To see this, note that we can find a simple closed curve α on the top sheet enclosing the slit from e_2 to e_3 and another closed curve β that goes along the top sheet from the first slit to the second, goes through the slits to the bottom sheet and closes up. These curves can be drawn so they intersect exactly once. They form generators for the fundamental group based at the intersection point.

The Weierstrass $\wp$-function satisfies a simple differential equation

$$\wp'^2 = 4\wp^3 - g_2\wp - g_3$$

where

$$g_2 = 60 \sum_{(m,n)\neq(0,0)} \omega_{m,n}^{-4} \text{ and } g_3 = 140 \sum_{(m,n)\neq(0,0)} \omega_{m,n}^{-6}.$$

To prove this, note that for z sufficiently close to the origin we have the expansion

$$\wp(z) = z^{-2} + \frac{1}{20}g_2 z^2 + \frac{1}{28}g_3 z^4 + O(z^6).$$

Differentiating, substituting and collecting terms we have

$$\wp'^2 - 4\wp^3 + g_2\wp + g_3 = O(z^2).$$

The left side is an elliptic function with the same periods as $\wp$; it is, however, holomorphic at the origin and therefore at all points congruent to the origin.

t cannot have poles anywhere else so it is holomorphic everywhere and thus constant. Letting $z \to 0$, we see the constant is zero.

Since we know the zeros of $\wp'$, we can write

$$\wp^2(z)' = 4(\wp(z) - e_1)(\wp(z) - e_2)(\wp(z) - e_3).$$

We then have $e_1 + e_2 + e_3 = 0$, $4(e_1 e_2 + e_2 e_3 + e_3 e_1) = -g_2$ and $4e_1 e_2 e_3 = g_3$.

Using this differential equation, one can show that any elliptic function with the same period lattice can be written as a rational function of $\wp$ and $\wp'$.

If we replace the period lattice by a similar lattice it is easy to see how the $\wp$-function transforms. From the definition, for any $k \in \mathbb{C}$, we have

$$\wp\left(\frac{z}{k}; \frac{a}{k}, \frac{b}{k}\right) = k^2 \wp(z; a, b).$$

It is often convenient to set $\tau = b/a$ and consider the $\wp$-function with period lattice generated by $(1, \tau)$. Then

$$\wp\left(\frac{z}{a}; 1, \tau\right) = a^2 \wp(z; a, b).$$

The coefficients g_2 and g_3 transform by

$$g_2 = g_2(a, b) = \frac{1}{a^4} g_2(1, \tau) \text{ and } g_3 = g_3(a, b) = \frac{1}{a^6} g_3(1, \tau).$$

The natural inverse question is whether, given constants g_2, g_3 such that the discriminant $g_2^3 - 27 g_3^2$ of the cubic expression does not vanish, there exists a solution of the differential equation

$$\left(\frac{dy}{dz}\right)^2 = 4y^3 - g_2 y - g_3$$

of the form

$$y = \beta(z; \pi, \pi\tau)$$

whose invariants are g_2 and g_3. This problem can be solved by the use of *theta functions*.

Theta functions have a rich theory in their own right and are the subject of many whole books. They were introduced and studied systematically by Jacobi. They are power series in the variables $q = e^{\pi i \tau}$ and z and they converge much more rapidly than the power series for the $\wp$-function. Here we give only the definitions and formulas we need to write down the solution to the inverse question. This is the solution we used in Chapter 14. We follow the standard notation found, for example, in [68], and the reader is referred there for proofs.

Set $q = e^{\pi i \tau}$ and define

$$\vartheta(z, q) = \vartheta_4(z, q) = \sum_{n=-\infty}^{\infty} (-1)^n q^{n^2} e^{2niz} = 1 + 2\sum_{n=1}^{\infty} (-1)^n q^{n^2} \cos 2nz.$$

Then

$$\vartheta_4(z + \pi, q) = \vartheta_4(z, q) \text{ and } \vartheta_4(z + \pi\tau, q) = -q^{-1} e^{-2iz} \vartheta_4(z, q).$$

Note that ϑ_4 is not doubly-periodic. It is, however, *quasi-doubly-periodic* because, as the above formulas show, it picks up a multiplier, 1 or $-q^{-1} e^{-2iz}$, on the addition of π or $\pi\tau$.

Next set

$$\vartheta_3(z, q) = \vartheta_4\left(z + \frac{\pi}{2}, q\right) = 1 + 2\sum_{n=1}^{\infty} q^{n^2} \cos 2nz,$$

$$\vartheta_1(z, q) = -ie^{iz + \frac{1}{4}\pi i \tau} \vartheta_4\left(z + \frac{\pi\tau}{2}, q\right) = 2\sum_{n=0}^{\infty} (-1)^n q^{(n+\frac{1}{2})^2} \sin(2n+1)z$$

and finally

$$\vartheta_2(z, q) = \vartheta_1\left(z + \frac{\pi}{2}, q\right) = 2\sum_{n=0}^{\infty} q^{(n+\frac{1}{2})^2} \cos(2n+1)z.$$

There are many interesting relations satisfied by these four theta functions. Setting $z = 0$, we obtain the *theta constants* which are actually functions of τ – hence the notation $\vartheta_i(0|\tau)$. Jacobi showed they have particularly nice product expansions in q; in particular,

$$\vartheta_3(0|\tau) = \Pi_{n=1}^{\infty} (1 - q^{2n})^2 (1 + q^{2n-1})^2$$

and

$$\vartheta_4(0|\tau) = \Pi_{n=1}^{\infty} (1 - q^{2n})^2 (1 - q^{2n-1})^2.$$

The inversion problem is equivalent to finding a holomorphic solution τ, $\Im\tau > 0$, to the equation

$$\frac{\vartheta_4(0|\tau)^4}{\vartheta_3(0|\tau)^4} = \Pi_{n=1}^{\infty} \left(\frac{(1 - q^{2n-1})}{(1 + q^{2n-1})}\right)^8. \tag{16.1}$$

This solution can be found by using Cauchy's integral formula and integrating around a fundamental domain for a Fuchsian group called the modular group. This is the group $PSL(2, \mathbb{Z})$. It is the group of transformations that preserve a given lattice, but change the generating periods. (See Example 1 in Section 6.3 and Exercises 6.1 and 6.8.)

Thus, given g_2 and g_3, let e_1, e_2, e_3 be the roots of the cubic. Label them so that the cross ratio

$$cr(e_1, e_3, \infty, e_2) = \frac{e_1 - e_3}{e_1 - e_2}$$

takes values in $\mathbb{C} \setminus ((-\infty, 0] \cup [1, \infty))$. Then, solving equation (16.1) for τ, we obtain, with $A = (e_1 - e_2)/\vartheta_4^4(0| \in)$, $\nu = Az$,

$$\wp(\nu; \pi, \pi\tau) = \frac{\vartheta_2 \nu(|\tau)^2}{\vartheta_1 \nu(|\tau)^2} \vartheta_3(0|\tau)^2 \vartheta_4(0|\tau)^2 + e_1. \qquad (16.2)$$

If the cross ratio falls into one of the forbidden intervals, the points can be re-labeled so that it doesn't (see Exercise 1.28).

The cross ratio function $cr(e_1, e_3, \infty, e_2)$ is thus a function of τ and is usually denoted by $\lambda(\tau)$. It maps the upper half plane onto the plane minus the forbidden intervals.

The generalization of the theory of elliptic functions to functions invariant under a Fuchsian group is called the theory of automorphic functions. Such functions take the same value at points congruent under the Fuchsian group G and thus define functions on the Riemann surface Δ/G. A good introduction to this theory can be found in [57]. See also [13] and [58] for extensive discussions.

Bibliography

[1] L. Ahlfors, *Lectures on Quasiconformal Mappings*, Van Nostrand Studies **10** (1966)

[2] L. Ahlfors, *Conformal Invariants* McGraw-Hill, 1973

[3] L. Ahlfors and L. Sario, *Riemann Surfaces*, Princeton Univ. Press, 1960

[4] I. N. Baker and P. J. Rippon, On compositions of analytic self-mappings of a convex domain, *Arch. Math. (Basel)* **55** (1990), no. 4, 380–386

[5] A. F. Beardon, *The Geometry of Discrete Groups*, Springer-Verlag, 1983

[6] A. F. Beardon, T. K. Carne, D. Minda and T. W. Ng, Random iteration of analytic maps, *Ergodic Th. and Dyn. Systems* **24** (2004), no. 3 659–675

[7] A. F. Beardon and C. Pommerenke, The Poincaré metric of plane domains, *J. London Math. Soc.* **18** (1978), 475–483

[8] P. Buser, The collar theorem and examples, *Manuscripta Math.* **25** (1978), 349–357

[9] L. Carleson and T. W. Gamelin, *Complex Dynamics*, Springer-Verlag, 1993

[10] M. Comerford, A survey of results in random iteration, *Fractal Geometry and Applications: A Jubilee of Benoit Mandelbrot. Proc. Sympos. Pure Math.* **72**, Part 1, 435–476, Amer. Math. Soc., 2004

[11] M. Comerford, Conjugacy and counterexample in random iteration. *Pacific J. Math.* **211** (2003), no. 1, 69–80

[12] S. Dineen, *The Schwarz Lemma*, Oxford Mathematical Monographs, Clarendon Press, Oxford University Press, 1989

[13] H. Farkas and I. Kra, *Riemann Surfaces*

[14] W. Fenchel, *Elementary Hyperbolic Geometry*, North-Holland

[15] L. Ford, *Automorphic Functions*, Chelsea

[16] J. E. Fornaess, Real methods in complex dynamics, *Real Methods in Complex and CR Geometry*, Lecture Notes in Mathematics **1848**, 49–107, Springer-Verlag, 2004

[17] F. P. Gardiner, *Teichmüller Theory and Quadratic Differentials*, Wiley-Interscience, 1987

[18] F. P. Gardiner and N. Lakic, *Quasiconformal Teichmüller Theory*, AMS Mathematical Surveys and Monographs **76**, Amer. Math. Soc., 2000

[19] F. P. Gardiner and N. Lakic, Comparing Poincaré distances, *Annals of Math.* **154** (2001), 245–267

20] J. Gill, Compositions of analytic functions on the form $F_n(z) = F_{n-1}(f_n(z))$, $f_n(z) \to f(z)$, *J. Comput. Appl. Math.* **23** (2), 1988, 179–184

21] R. Harmelin and D. Minda, Quasi-invariant domain constants, *Israel J. Math.* **77** (1992), no. 1–2, 115–127

22] J. A. Hempel, The Poincaré metric on the twice punctured plane and the theorems of Landau and Schottky, *J. London Math. Soc.* (2) **20** (1979), 435–445

23] E. Hille, *Analytic Function Theory* Vols. I and II, Ginn, 1962

24] A. Hinkkanen, Julia sets of rational functions are uniformly perfect, *Math. Proc. Cambridge Philos. Soc.* **113** (1993), no. 3, 543–559

25] A. Hinkkanen and G. J. Martin, Julia sets of rational semigroups, *Math. Z.* **222** (1996), no. 2, 161–169

26] P. Järvi and M. Vuorinen, Uniformly perfect sets and quasiregular mappings, *J. London Math. Soc.* (2) **54** (1996), no. 3, 515–529

27] M. Kapovich, *Hyperbolic Manifolds*, Birkhäuser, 2001

28] L. Keen, Canonical polygons for finitely generated Fuchsian groups, *Acta Math.* **115** (1966) 1–16

29] L. Keen, Intrinsic moduli on Riemann surfaces, *Annals of Math.* **84** (3) (1966), 404–420

30] L. Keen, Collars on Riemann Surfaces, in *Discontinuous Groups and Riemann Surfaces*, 263–268, *Annals of Math. Studies* **79**, Princeton Univ. Press (1974)

31] L. Keen and N. Lakic, Random holomorphic iterations and degenerate subdomains of the unit disk, *Proc. Amer. Math. Soc.* **134** (2) (2006), 371–378

32] L. Keen and N. Lakic, Forward iterated function systems, *Complex Dynamics and Related Topics, Lectures at the Morningside Center of Mathematics*, New Studies in Advanced Mathematics, IP Vol 5 2003

33] L. Keen and N. Lakic, Accumulation points of iterated function systems, in *Complex Dynamics: Twenty-Five Years after the Appearance of the Mandelbrot Set*, Contemp. Math., Eds. R. Devaney and L. Keen, Ame. Math. Soc., 2006

34] L. Keen and N. Lakic, Accumulation constants of iterated function systems with Bloch target domains, *Annales Acad. Sci. Fenn.* **32** Ser A. J Math, 1–10 (2007)

35] L. Keen and N. Lakic, A generalized hyperbolic metric for plane domains submitted

36] S. G. Krantz, *Complex Analysis: A Geometric Viewpoint*, Carus Mathematic Monographs, **23**, MAA, 1990

37] L. Lorentzen, Compositions of contractions, *J. Comput. Appl. Math.* **32** (1990), 169–178

38] W. Ma, F. Maitani and D. Minda, Two-point comparisons between hyperbolic and Euclidean geometry on plane regions. Dedicated to Professor Eligiusz J. Złotkiewicz, *Ann. Univ. Mariae Curie-Skłodowska Sect. A* **52** (1998), no. 1, 83–96

39] B. Maskit, *Kleinian Groups*, Springer-Verlag, 1988

40] B. Maskit, Comparison of hyperbolic and extremal lengths. *Ann. Acad. Sci. Fenn. Ser. A I Math.* **10** (1985), 381–386

41] W. Massey, *Algebraic Topology: An Introduction*, Harcourt-Brace, 1967

42] C. T. McMullen, *Complex Dynamics and Renormalizations*, *Annals of Math. Studies* **135**, Princeton Univ. Press, 1994

[43] J. Milnor, *Dynamics in One Complex Variable: Introductory Lectures*, 3rd Ed., Princeton Univ. Press, 2006

[44] D. Minda, Bloch and normal functions on general planar regions, *Holomorphic Functions and Moduli*, Vol. I (Berkeley, CA, 1986), 101–110, Mathematical Sciences Research Institute Publications **10**, Springer-Verlag, 1988

[45] R. Mañé and L. F. da Rocha, Julia sets are uniformly perfect *Proc. Amer. Math. Soc.* **116** (1992), no. 1, 251–257

[46] K. Matsuzaki and M. Taniguchi, *Hyperbolic Manifolds and Kleinian Groups*, Oxford Mathematical Monographs, Oxford Univ. Press, 1998

[47] D. Mauldin, F. Przytycki and M. Urbański, Rigidity of conformal iterated function systems, *Compositio Math.* **129** (2001), 273–299

[48] V. Mayer, D. Mauldin and M. Urbański, Rigidity of connected limit sets of iterated function systems, *Mich. Math J.* **49** (2001), 451–458

[49] J. R. Munkres, *Topology*, 2nd Ed. Prentice-Hall, 2000

[50] Z. Nehari *Conformal Mapping*, McGraw-Hill, 1952

[51] Z. Nehari *Analytic Functions*, Springer-Verlag, 1970

[52] B. Osgood, Some properties of $\frac{f''}{f'}$ and the Poincaré metric, *Indiana Univ. Math. J.* **31** (1982), 449–461

[53] H. Poincaré, *Theory of Fuchsian Groups, Papers on Fuchsian Functions* (John Stillwell, translator) Springer-Verlag, 1985

[54] H. Poincaré, *Science and Method* (Francis Maitland, translator), Dover, 1952

[55] C. Pommerenke, Uniformly perfect sets and the Poincaré metric, *Arch. der Math.* **32** (1979), 192–199

[56] F. Ren, J. Zhou and W. Qiu, Self-similar measures on the Julia sets. *Progr. Natur. Sci. (English Ed.)* **8** (1998), no. 1, 24–34

[57] C. L. Siegel, *Topics in Complex Function Theory*, volume II, Interscience Tracts in Pure and Applied Mathematics, Number 25, wiley

[58] G. Springer, *Riemann Surfaces*, Chelsea, 1981

[59] S. Stahl, *The Poincaré Half-Plane*, Jones and Bartlett, 1993

[60] B. Solomyak and M. Urbański, L^q densities for measures associated with parabolic iterated function systems with overlaps, *Indiana J. Math.* **50** 2001

[61] R. Stankewitz, Uniformly perfect analytic and conformal attractor sets. *Bull. London Math. Soc.* **33** (2001), no. 3, 320–330

[62] R. Stankewitz, Uniformly perfect sets, rational semigroups, Kleinian groups and IFS's, *Proc. Amer. Math. Soc.* **128** (2000), no. 9, 2569–2575

[63] T. Sugawa, Uniformly perfect sets: analytic and geometric aspects, *Sugaku Expo.* **16** (2003), 225–242

[64] T. Sugawa and M. Vourinen, Some inequalities for the Poincaré metric of plane domains, *Math. Z.* **250** (2005), 885–906

[65] M. Urbański, A. Zdunik, Hausdorff dimension of harmonic measure for self-conformal sets, *Adv. Math.* **171** (2002), no. 1, 1–58

[66] W. Veech *A Second Course in Complex Analysis*, Benjamin, 1967

[67] S. Wang and L.-W. Liao, Uniformly perfect Julia sets of meromorphic functions, *Bull. Austral. Math. Soc.* **71** (2005), no. 3, 387–397

[68] E. T. Whittaker and G. N. Watson, *A Course of Modern Analysis*, Cambridge Univ. Press, 1996

69] K. Włodarczyk, D. Klim and E. Gontarek, Random iterations of holomorphic maps in complex Banach spaces, *Proc. Amer. Math. Soc.* **128** (2000), no. 12, 3475–3482

70] K. Włodarczyk, D. Klim and E. Gontarek, Random iterations, fixed points and invariant CRF-horospheres in complex Banach spaces, *J. Math. Anal. Appl.* **295** (2004), no. 2, 291–302

Index

Printed in the United States
by Baker & Taylor Publisher Services